Essential Results of Functional Analysis

/

Chicago Lectures in Mathematics Series
J. Peter May, Robert J. Zimmer, Spencer J. Bloch,
and Norman R. Lebovitz, editors

Other *Chicago Lectures in Mathematics* titles
available from The University of Chicago Press:

The Theory of Sheaves, by Richard G. Swan (1964)
Topics in Ring Theory, by I. N. Herstein (1969)
Fields and Rings, by Irving Kaplansky (1969; 2d ed. 1972)
Infinite Abelian Group Theory, by Phillip A. Griffith (1970)
Topics in Operator Theory, by Richard Beals (1971)
Lie Algebras and Locally Compact Groups, by Irving Kaplansky (1971)
Several Complex Variables, by Raghavan Narasimhan (1971)
Torsion-Free Modules, by Eben Matlis (1973)
The Theory of Bernoulli Shifts, by Paul C. Shields (1973)
Stable Homotopy and Generalized Homology, by J. F. Adams (1974)
Commutative Rings, by Irving Kaplansky (1974)
Banach Algebras, by Richard Mosak (1975)
Rings with Involution, by I. N. Herstein (1976)
Theory of Unitary Group Representation, by George W. Mackey (1976)
Infinite-Dimensional Optimization and Convexity, by Ivar Ekeland and
   Thomas Turnbull (1963)
Commutative Semigroup Rings, by Robert Gilmer (1984)
Navier-Stokes Equations, by Peter Constantin and
   Ciprian Foias (1988)

Robert J. Zimmer

Essential Results of Functional Analysis

The University of Chicago Press
Chicago and London

ROBERT J. ZIMMER is professor of mathematics at the University of Chicago. He is the author of *Ergodic Theory and Semisimple Groups.*

The University of Chicago Press, Chicago 60637
The University of Chicago Press, Ltd., London

© 1990 by The University of Chicago
All rights reserved. Published 1990
Printed in the United States of America

99 98 97 96 95 94 93 92 91 90   54321

Library of Congress Cataloging-in-Publication Data

Zimmer, Robert J., 1947–
    Essential results of functional analysis / Robert J. Zimmer.
        p.   cm.—(Chicago lectures in mathematics series)
    Includes bibliographical references.
    ISBN 0-226-98337-4. — ISBN 0-226-98338-2 (pbk.)
    1. Functional analysis.   I. Title.   II. Series.
QA320.Z58   1990
515—dc20   89-28992
                                                       CIP

The paper used in this publication meets the minimum requirements of the American National Standard for Information Sciences—Permanence of Paper for Printed Library Materials, ANSI Z39.48-1984.

To Terese, David, and Benjamin

TABLE OF CONTENTS

PREFACE

These notes are based upon a one-quarter course I gave for first year graduate students at the University of Chicago in 1985 and again in 1986. The aim of this course was to introduce the students to substantive and important results in a diversity of areas within analysis to which functional analysis makes an important contribution, and to demonstrate the unity of perspective and technique that the functional analytic approach offers. The course assumed a basic knowledge of measure theory and the elementary theory of Banach and Hilbert spaces. (In Chicago, this material is covered in the preceding quarter. There are, of course, many satisfactory expositions of this material in the literature.) In these notes, we have summarized this necessary background information in Chapter 0. In the remaining chapters, whose general content the reader will discover in the table of contents, our approach was to focus upon central theorems; thus, we do not develop all possible related machinery and foundations concerning a given subject, but enough framework and examples to be able to present these theorems in an understandable yet concise manner. Most of the material in the first five chapters was covered in the quarter; the sixth chapter I added because the notes felt unfinished without it. A natural alternative for a quarter would be to cover chapters 1-3 and 5-6. One should be able to cover all six chapters in a semester.

The task of writing these notes was greatly facilitated by the lecture notes taken by Paul Burchard during the course. He prepared them with great care and made a number of improvements in the exposition. The task of presenting the lectures was greatly facilitated by the interest, effort, and enthusiasm of the graduate students at the University of Chicago who attended the course. To Paul Burchard, in particular, and to the class in general, I wish to express my appreciation.

BACKGROUND

0.A. Review of Basic Functional Analysis

We review in this preliminary chapter some basic definitions, results, and examples concerning Banach spaces.

We let k be the field $\mathbf{R}$ or $\mathbf{C}$, and E a vector space over k.

DEFINITION A.1. *A norm on E is a map $\| \ \| : E \to \mathbf{R}$ such that*

    (i) $\|x\| \geq 0$ and $\|x\| = 0$ if and only if $x = 0$.

    (ii) *For $c \in k$ and $x \in E$, $\|cx\| = |c| \|x\|$.*

    (iii) $\|x + y\| \leq \|x\| + \|y\|$. *(triangle inequality)*

If E is endowed with a given norm we call E a normed linear (or vector) space.

EXAMPLE A.2: (a) Let X be a compact space and $C(X)$ the space of k-valued continuous functions on X. For $f \in X$, define $\|f\| = \sup\{|f(x)| \, | \, x \in X\}$. We remark that $\|f\| < \infty$ since X is compact. (b) More generally, let X be any Hausdorff space and let $BC(X)$ be the space of bounded continuous functions on X. Then defining $\|f\|$ as in (a), $BC(X)$ has the structure of a normed vector space. (c) On k^n, we have a variety of norms. For example, for $x = (x_1, \ldots, x_n) \in k^n$, $\|x\|_p = \left\{ \sum |x_i|^p \right\}^{1/p}$ is a norm, as is $\|x\|_\infty = \sup\{|x_i|\}$. We remark that with $\| \ \|_\infty$, we have $k^n = C(\{1, \ldots, n\})$, where $C(X)$ is as in (a).

We now recall the basic normed spaces of measure theory. Throughout these notes, by "measure" we *always* mean a positive σ-finite measure. Now let (X, μ) be a measure space. If $f \colon X \to k$ is measurable, and $1 \leq p < \infty$, we set $\||f|\|_p = (\int_X |f|^p d\mu)^{1/p}$. We let $\mathcal{L}^p(X) = \{f \, | \, \||f|\|_p < \infty\}$. The map $f \mapsto \||f|\|_p$ is not a norm on the vector space $\mathcal{L}^p(X)$. Namely, if $f = 0$ a.e., but f is not identically 0, then we have $\||f|\|_p = 0$ for a nonzero f. However, $\|| \ |\|_p$ is a semi-norm in the following sense.

DEFINITION A.3. *If E is a vector space over k, a semi-norm on E is a map $\| \ \|: E \to \mathbf{R}$ such that:*

(i) $\|x\| \geq 0$
(ii) $\|cx\| = |c|\,\|x\|$ *for* $c \in k, x \in E$.
(iii) $\|x + y\| \leq \|x\| + v\|y\|$.

EXAMPLE A.4: Any semi-norm $\| \| \ \| \|$ on a vector space E yields a normed linear space in the following way. Let $E_0 = \{x \in E \mid \| \| x \| \| = 0\}$. Then $E_0 \subset E$ is a linear subspace and the map $\| \| \ \| \|: E \to \mathbf{R}$ factors to a map $\| \ \|: E/E_0 \to \mathbf{R}$, which is a norm on the space E/E_0. For the seminorms $\| \| \ \| \|_p$ on $\mathcal{L}^p(X)$ defined above, we denote the corresponding normed space by $L^p(X)$ and the norm by $\| \ \|_p$. In this case $E_0 = \{f: X \to k \mid f$ is measurable and $f = 0$ a.e.$\}$. Thus, we can view $L^p(X)$ as the space of measurable functions on X such that $\int |f|^p < \infty$, with two such functions being identified if they agree a.e.

Example A.4 represents one use of the notion of a seminorm. However seminorms shall play a basic role in later developments in these notes.

EXAMPLE A.5: If (X, μ) is a measure space, Y is a topological space, and $f: X \to Y$ is measurable, we recall that $y \in Y$ is said to be in the essential range of f if for any open neighborhood U of y, $\mu(f^{-1}(U)) > 0$. If $Y = k$, then f is called essentially bounded if the essential range is a bounded set, and we let $\mathcal{L}^\infty(X) = \{f: X \to k \mid f$ is essentially bounded$\}$. We define $\| \| f \| \|_\infty = \sup\{|z| \mid z$ is in the essential range of $f\}$. As in the previous examples $\| \| \ \| \|_\infty$ is a semi-norm, and the corresponding normed space and norm are denoted by $L^\infty(X), \| \ \|_\infty$.

We remark that one can define $L^p(X)$ for $0 < p < 1$ as well, however in that case $\| \ \|_p$ does not satisfy the triangle inequality.

If E is a normed linear space, we define a metric on E by $d(x, y) = \|x - y\|$.

DEFINITION A.6. *If E is a normed linear space, E is called a Banach space if E is complete with respect to the metric defined by the norm.*

EXAMPLE A.7: (a) If X is a Hausdorff space, $BC(X)$ is complete. In particular, if X is compact $C(X)$ is complete.

(b) (Riesz-Fischer theorem). If (X, μ) is a measure space, $L^p(X)$ is complete for $1 \leq p \leq \infty$.

It is often useful to know when a subspace of a normed space is dense.

THEOREM A.8. (Stone-Weierstrass) *Let X be a compact Hausdorff space, and $\mathcal{A} \subset C(X)$ be a linear subspace. Suppose further that*

- (i) *$\mathcal{A}$ is a subalgebra, i.e. is closed under multiplication.*
- (ii) *$\mathcal{A}$ contains the constant functions.*
- (iii) *$\mathcal{A}$ separates points; i.e., for $x, y \in X$, there is some $f \in \mathcal{A}$ such that $f(x) \neq f(y)$.*
- (iv) *In case $k = \mathbf{C}, f \in \mathcal{A}$ implies $\overline{f} \in \mathcal{A}$.*

Then $\mathcal{A}$ is dense in $C(X)$.

EXAMPLE A.9: (a) Suppose $X \subset \mathbf{R}^n$ is a compact set. Let $\mathcal{P}(X) \subset C(X)$ be the restrictions of polynomial functions on $\mathbf{R}^n$ to X. Then A.8 applies and $\mathcal{P}(X)$ is dense in $C(X)$.
(b) Let $S^1 \subset \mathbf{C}$ be the unit circle. Let $P(S^1)$ be the set of functions of the form $f(z) = \sum_{n=-N}^{N} a_n z^n$ where N is an arbitrary finite natural number. Then $P(S^1)$ is dense in $C(S^1)$.
(c) Let X be a compact Hausdorff space and $\mathcal{A} \subset C(X \times X)$ be the set of all finite linear combinations of functions of the form $(x, y) \rightarrow f(x)g(y)$ where $f, g \in C(X)$. Then $\mathcal{A}$ is dense in $C(X \times X)$.

PROPOSITION A.10. *Let X be a locally compact separable metric space and μ a measure on X which is finite on compact sets. Let $C_c(X) \subset C(X)$ be the set of compactly supported continuous functions. Then for any $p, 1 \leq p < \infty, C_c(X)$ is dense in $L^p(X)$.*

A.10 follows directly from "regularity" of the measure.
For any open set $U \subset \mathbf{R}^n$, we shall always suppose U is endowed with Lebesgue measure, unless we specifically state otherwise.

PROPOSITION A.11. *Let $U \subset \mathbf{R}^n$ be an open set and $C_c^\infty(U)$ be the space of smooth functions with compact support contained in*

U, i.e. $support(f) \subset U$. Then for $1 \leq p < \infty, C_c^\infty(U)$ is dense in $L^p(U)$. (For a proof, see B.6 below.)

A map $T: E \rightarrow F$ between normed spaces is continuous if it is continuous with respect to the topologies defined by the metrics on E, F. A linear map $T: E \rightarrow F$ is called bounded if there is a number B such that $\|Tx\| \leq B\|x\|$ for all $x \in E$. Equivalently, if we set $E_r = \{x \in E \mid \|x\| \leq r\}$, then $T(E_1) \subset F_B$.

PROPOSITION A.12. *If E, F are normed spaces, and $T: E \rightarrow F$ is a linear map, then the following are equivalent:*

    (a)   *T is continuous.*
    (b)   *T is bounded.*
    (c)   *T is continuous at $0 \in E$.*

DEFINITION A.13. *(i) A linear bijection $T: E \rightarrow F$ is called an isomorphism of the normed spaces E, F if T and T^{-1} are continuous. A linear map $T: E \rightarrow F$ is called an isometry if $\|Tx\| = \|x\|$ for all $x \in E$. T is called an isometric isomorphism if it is both an isometry and an isomorphism. Equivalently, T is a bijective isometry.*
(ii) If $\| \; \|_1$ and $\| \; \|_2$ are norms on a vector space E, $\| \; \|_1$ and $\| \; \|_2$ are called equivalent if the identity map $(E, \| \; \|_1) \rightarrow (E, \| \; \|_2)$ is an isomorphism. By A.12, this is equivalent to the existence of a constants $b, c > 0$ such that

$$b\|x\|_1 \leq \|x\|_2 \leq c\|x\|_1$$

for all $x \in E$.

In finite dimensions, any bijective continuous linear map T is automatically an isomorphism, since all linear maps (and in particular T^{-1}) are continuous. For Banach spaces we have:

PROPOSITION A.14. *(Open mapping theorem). If $T: E \rightarrow F$ is a continuous bijection of Banach spaces, then T is an isomorphism. (Equivalently, T is an open map.)*

We let $B(E, F)$ denote the space of bounded linear maps from E to F. It is clearly a vector space, and becomes a normed linear space if we define, for $T \in B(E, F)$, $\|T\| = \sup\{\|Tx\| \mid \|x\| \leq 1\}$.

PROPOSITION A.15. *If F is a Banach space, so is $B(E, F)$.*

EXAMPLE A.16: If E is a normed linear space, we denote $B(E, k)$ by E^*, and $B(E, E)$ by $B(E)$. Then E^* is always a Banach space and $B(E)$ will be a Banach space if E is Banach.

EXAMPLE A.17: If $\dim E < \infty$, so that $E \cong k^n$ as a vector space, then any linear functional (i.e. linear map into k) on E is continuous. Thus, $E^* \cong k^n$ as a vector space. The norm on E^* of course depends upon the norm on E.

If X is a measure space, $f \in L^p(X)$ and $h \in L^q(X)$, then $fh \in L^1(X)$ as long as p and q are related by $\dfrac{1}{p} + \dfrac{1}{q} = 1$. (For $p = 1$, we let $q = \infty$). In fact, we have the basic Hölder inequality: $\|fh\|_1 \leq \|f\|_p \|h\|_q$. Fix $1 \leq p \leq \infty$. For $h \in L^q(X)$ let $\lambda_h \in L^p(X)^*$ be given by $\lambda_h(f) = \int fh$.

PROPOSITION A.18. *The map $L^q(X) \to L^p(X)^*, h \mapsto \lambda_h$, is an isometric isomorphism for $1 \leq p < \infty$. For $p = \infty$ this is an (injective) isometry, but is not in general an isomorphism.*

Suppose now that X is a compact metric space. Let $M(X)$ denote the space of probability measures on X, i.e. measures with $\mu(X) = 1$. For $\mu \in M(X)$, define $\lambda_\mu \in C(X)^*$ by $\lambda_\mu(f) = \int f d\mu$.

THEOREM A.19. (Riesz representation theorem). *The map $M(X) \to C(X)^*, \mu \mapsto \lambda_\mu$, is a bijection of $M(X)$ with $\{\lambda \in C(X)^* \mid \lambda(f) \geq 0$ for all $f \geq 0, \lambda(1) = 1\}$.*

COROLLARY A.20. *For $k = \mathbf{R}$, every $\lambda \in C(X)^*$ can be written uniquely as $\lambda_{\mu_1} - \lambda_{\mu_2}$ where μ_1, μ_2 are finite measures. For $k = \mathbf{C}$, every $\lambda \in C(X)^*$ can be written uniquely as $(\lambda_{\mu_1} - \lambda_{\mu_2}) + i(\lambda_{\mu_3} - \lambda_{\mu_4})$ where $\mu_i, 1 \leq i \leq 4$, are finite measures.*

In all the above examples it is easy to see that for any $x \in E, x \neq 0$, there is some $\lambda \in E^*$ with $\lambda(x) \neq 0$. This holds in general.

THEOREM A.21. (Hahn-Banach) *Let E be a normed linear space and $F \subset E$ a linear subspace. For any $\lambda \in F^*$, there is some $\tilde{\lambda} \in E^*$ such that $\tilde{\lambda}\big| F = \lambda$ and $\|\tilde{\lambda}\| = \|\lambda\|$.*

COROLLARY A.22. *If $x \in E$, then there exists some $\lambda \in E^*$ with $\|\lambda\| = 1$ and $|\lambda(x)| = \|x\|$.*

REMARK A.23: The proof of the Hahn-Banach theorem works for spaces with a semi-norm, not just a norm. As we shall need this later on, we state this version now. Let E be a linear space with a semi-norm $\|\ \|$. Let $F \subset E$ be a linear subspace. Suppose $\lambda: F \to k$ is linear and $B > 0$ such that $|\lambda(x)| \leq B\|x\|$ for all $x \in F$. Then there is a linear map $\tilde{\lambda}: E \to k$ such that $\tilde{\lambda}\big| F = \lambda$ and $|\lambda(x)| \leq B\|x\|$ for all $x \in E$.

DEFINITION A.24. *Let E be a vector space over k. An inner product on E is a map $\langle\ ,\ \rangle : E \times E \to k$ such that*
  a) $\langle\ ,\ \rangle$ *is bilinear for $h = \mathbb{R}$; or for $h = \mathbb{C}$ is bilinear over $\mathbb{R}$, linear over $\mathbb{C}$ in the first variable, and satisfies $\langle x, iy \rangle = -i \langle x, y \rangle$.*
  b) $\langle x, x \rangle \geq 0$ *and $\langle x, x \rangle = 0$ if and only if $x = 0$.*

EXAMPLE A.25: (a) For $E = k^n$, $z = (z_1, \ldots, z_n)$, $w = (w_1, \ldots, w_n) \in k^n$, let $\langle z, w \rangle = \sum_{i=1}^{n} z_i \overline{w}_i$.
(b) For $E = L^2(X)$, let $\langle f, g \rangle = \int f\overline{g}$.

Any inner product on E defines a norm on E by $\|x\| = \langle x, x \rangle^{1/2}$, and thus E is also a normed linear space. A complete inner product space is called a Hilbert space. Both k^n and $L^2(X)$ are thus Hilbert spaces. We remark that k^n is identical to $L^2(\{1, \ldots, n\})$ with counting measure.

A subset $A \subset E$, where E is a Hilbert space is called orthonormal if $\|x\| = 1$ for all $x \in A$ and $x, y \in A$, $x \neq y$ implies $x \perp y$, i.e. $\langle x, y \rangle = 0$. A is called a maximal orthonormal set if it is not a proper subset of an orthonormal set. Every orthonormal subset of a Hilbert space is contained in a maximal orthonormal set by Zorn's lemma. A maximal orthonormal set is also called an orthonormal basis.

PROPOSITION A.26. *Let E be a Hilbert space and $A \subset E$ an orthonormal set. Then the following are equivalent:*
  (i) A *is maximal.*
  (ii) $y \in E$, $x \perp y$ for all $x \in A$ implies $y = 0$.

(iii) *For any $y \in E$, y can be uniquely expressed as $y = \sum_{x \in A} c_x \cdot x$ where $c_x \in k$ and $\sum |c_x|^2 < \infty$. Furthermore every such sum converges, and $c_x = \langle y, x \rangle$.*

(iv) *Finite linear combinations of elements of A are dense in E.*

COROLLARY A.27. *Let E be a Hilbert space. The map $E \to E^*$ given by $x \mapsto \lambda_x$, where $\lambda_x(y) = \langle y, x \rangle$ is a norm preserving bijection, preserving addition. For $h = \mathbf{R}$, this is linear; for $h = \mathbf{C}$, $\lambda_{cx} = \bar{c}\lambda_x$.*

COROLLARY A.28. *Let $W \in E$ be a closed linear subspace of a Hilbert space. Let $W^\perp = \{y \in E \mid \langle x, y \rangle = 0 \text{ for all } x \in W\}$. Then $W^\perp$ is a closed linear subspace such that $W \oplus W^\perp = E$.*

EXAMPLE A.29: (L^2-Fourier series) Let S^1 be the circle with the measure μ = normalized arc length, i.e. μ = (arc length)/2π. For each $n, -\infty < n < \infty$, let $f_n(z) = z^n$. Then $\{f_n\}$ is orthonormal by direct computation, and is maximal orthonormal by Example A.9.a and Proposition A.26. Thus for every $f \in L^2(S^1)$, we have $f = \sum_{-\infty}^{\infty} \langle f_n, f \rangle f_n$. (Here of course the infinite sum is convergent in L^2. Other types of convergence, with varying hypotheses on f, are more delicate.)

DEFINITION A.30. *A normed linear space is called separable if it is separable as a metric space, i.e. it has a countable dense subset.*

EXAMPLE A.31: (a) If X is a compact metric space, $C(X)$ is separable.
(b) If X is a separable metric space and μ is a measure on X, then for $1 \le p < \infty$, $L^p(X)$ is separable. In general, $L^\infty(X)$ is not separable.
(c) A Hilbert space is separable if and only if there is a countable maximal orthonormal set. In this case every maximal orthonormal set is countable.

We conclude this review by recalling that every metric space E has a "completion" $\overline{E}$. If E is a normed linear space, then $\overline{E}$ will be a Banach space in which E is isometrically embedded as a dense subspace. If E is an inner product space, $\overline{E}$ will be a Hilbert space. If E is separable, $\overline{E}$ will be separable as well.

0.B. Some special properties of integration in $\mathbf{R}^n$

We collect here some basic properties concerning integration in $\mathbf{R}^n$ that are not features of general measure spaces. Unlike the review in section A, we shall here provide complete proofs. The first result concerns differentiation under an integral. For $U \subset \mathbf{R}^n$ open, we let, as usual, $C^k(U)$ be the space of k-times continuously differentiable functions on U, and $C_c^k(U)$ those functions in $C^k(U)$ that have compact support, and this support is contained in U.

LEMMA B.1. *Let X be a measure space, $I \subset \mathbf{R}$ open and $f : I \times X \to \mathbf{C}$ a measurable function. Suppose that for each $x \in X$, $f(\cdot, x) \in C^1(I)$ and that for each $t \in I$, $f(t, \cdot), \frac{\partial f}{\partial t}(t, \cdot) \in L^1(X)$. Define $F(t) = \int_X f(t, x)dx$ and $G(t) = \int_X \frac{\partial f}{\partial t}(t, x)dx$. If G is continuous, then F is C^1 and $F'(t) = G(t)$.*

PROOF: Fix $t \in I$ and let $\varepsilon > 0$. Choose δ such that $|s| < \delta$ implies $|G(t + s) - G(t)| < \varepsilon$. Then for $|h| < \delta$, we have

$$\left| \frac{F(t+h) - F(t)}{h} - G(t) \right|$$

$$= \left| \int_X \left(\frac{f(t+h, x) - f(t, x)}{h} - \frac{\partial f}{\partial t}(t, x) \right) dx \right|$$

$$= \left| \int_X \left(\frac{1}{h} \int_0^h \frac{\partial f}{\partial t}(t+s, x)ds - \frac{\partial f}{\partial t}(t, x) \right) dx \right|$$

$$= \left| \frac{1}{h} \int_0^h \left(\int_X \frac{\partial f}{\partial t}(t+s, x)dx \right) ds - \int_X \frac{\partial f}{\partial t}(t, x)dx \right| \quad \text{(by Fubini}$$

$$= |\frac{1}{h}(\int_0^h [\int_X \frac{\partial f}{\partial t}(t+s,x)dx - \int_X \frac{\partial f}{\partial t}(t,x)dx]ds)|$$

$$= |\frac{1}{h}\int_0^h [G(t+s) - G(t)]ds|$$

$$\le \varepsilon.$$

Thus $F'(t) = G(t)$.

We now wish to show that $C_c^\infty(\Omega)$ is dense in $L^p(\Omega)$ for any open $\Omega \subset \mathbf{R}^n$. This is a consequence of an extremely useful general "smoothing" procedure that we now describe. Suppose $\delta\colon \mathbf{R}^n \to \mathbf{R}$ is measurable with $\delta \ge 0$ and $\int \delta = 1$. Then $\delta(t)dt$ is of course a probability measure on $\mathbf{R}^n$, as is $\delta(t-x)dt$ for any $x \in \mathbf{R}^n$. The measure $\delta(t-x)dt$ is just the "translation" of $\delta(t)dt$ by x. Given another function f, we consider the function $x \mapsto \int f(t)\delta(t-x)\,dt$, when this is defined. This function, which of course depends upon δ and f, has at x the value which is simply the average of f with respect to the probability measure $\delta(t-x)dt$. It is algebraically a bit more convenient to consider the function defined this way by $\widetilde{\delta}$ and f, where $\widetilde{\delta}(x) = \delta(-x)$. Thus, we define the convolution of δ and f by $(\delta * f)(x) = \int f(t)\delta(x-t)\,dt$ (when this integral is defined). Of course, if $\delta(x) = \delta(-x)$, as will often be the case in our applications, $(\delta * f)(x)$ agrees with the expression above. This definition has the advantage that $(\delta * f)(x) = (f * \delta)(x)$, as one sees immediately from the change of variables $t' = x - t$. From the definition we see that if δ is supported on $\{t \mid \|t\| \le \varepsilon\}$, then $(\delta * f)(x)$ will be a weighted average of the values of f over $\{y \mid \|y - x\| \le \varepsilon\}$. To illustrate how to obtain approximations to f by this procedure, suppose for the moment that $f \in C_c(\mathbf{R}^n)$. Then f is uniformly continuous. Hence for any $r > 0$ we can find $\varepsilon > 0$ so that $\|y - x\| \le \varepsilon$ implies $|f(y) - f(x)| \le r$. If δ is supported on $\{t \mid \|t\| \le \varepsilon\}$, it follows that for any x,

$$(\delta * f)(x) = \int_{\mathbf{R}^n} f(t)[\delta(x-t)dt] = \int_{\|t-x\|\le\varepsilon} f(t)[\delta(x-t)dt]$$

Since $|f(t) - f(x)| \leq r$ for all such t and $\delta(x - t)dt$ is a probability measure, we have $|(\delta * f)(x) - f(x)| \leq r$ as well. Hence, by shrinking the support of δ we can approximate f uniformly by functions of the form $\delta * f$.

We now discuss this more formally and obtain approximations in L^p.

DEFINITION B.2. *If δ, f are measurable functions on $\mathbf{R}^n$, we set*

$$(\delta * f)(x) = \int_{t \in \mathbf{R}^n} f(t)\delta(x - t)dt$$

whenever the integral is absolutely convergent.

PROPOSTION B.3.
(i) *If $\delta \in L^1(\mathbf{R}^n)$ and $f \in L^p(\mathbf{R}^n)$, then*

$$\delta * f \in L^p(\mathbf{R}^n) \text{ and } \|\delta * f\|_p \leq \|\delta\|_1 \|f\|_p.$$

(ii) *If $\delta \in C_c^\infty(\mathbf{R}^n)$ and $f \in L^p(\mathbf{R}^n)$, then*

$$\delta * f \in C^\infty(\mathbf{R}^n) \quad \text{and} \quad \frac{\partial}{\partial x_i}(\delta * f) = \frac{\partial}{\partial x_i}(\delta) * f.$$

(iii) *If $\delta, f \in C_c^\infty(\mathbf{R}^n)$ then*

$$\frac{\partial}{\partial x_i}(\delta * f) = \delta * \frac{\partial f}{\partial x_i},$$

(iv) *If $\delta \in L^1, f \in L^p, h \in L^q$ $(1 \leq p \leq \infty, \frac{1}{p} + \frac{1}{q} = 1)$ then*

$$(\delta * f, h) = (f, \delta^* * h) \quad \text{where } \delta^*(x) = \overline{\delta(-x)} \text{ (and } (f, h) = \int f\bar{h}).$$

PROOF: (i) This is clear for $p = \infty$. For $p < \infty$, let $h \in L^q$ where $1/p + 1/q = 1$. It suffices to see that $(\delta * f) \cdot h \in L^1(\mathbf{R}^n)$ and

$\|(\delta * f)h\|_1 \leq \|\delta\|_1 \|f\|_p \|h\|_q$, by virtue of A.18. Since $\delta * f = f * \delta$, we have

$$\int |(\delta * f)(x)h(x)|dx \leq \int\int |f(x - t)\delta(t)h(x)|dtdx$$

$$\leq \int |\delta(t)|\left(\int |f(x - t)||h(x)|dx\right)dt$$

Since

$$\int |f(x)|^p = \int |f(x - t)|^p,$$

we obtain

$$\|(\delta * f)h\|_1 \leq \int |\delta(t)| \, \|f\|_p \|h\|_q \, dt \leq \|\delta\|_1 \|f\|_p \|h\|_q.$$

(ii) Choose R so that $\mathrm{supp}(\delta) \subset \{t \mid \|t\| \leq R\}$. Hence, for any x,

$$(\delta * f)(x) = \int_{\|t\| \leq R + \|x\|} f(t)\delta(x - t)dt.$$

Thus, if we fix x_0, then for all x with $\|x - x_0\| \leq 1$, we have

$$(\delta * f)(x) = \int \chi(t)f(t)\delta(t - x)dt$$

where $\chi(t)$ is the characteristic function of

$$A = \{t \mid \|t\| \leq R + \|x_0\| + 1\}.$$

I.e., for $\|x - x_0\| \leq 1$, $(\delta * f)(x) = (\delta * \tilde{f})(x)$ where $\tilde{f} = \chi f$. The point of this is that A is bounded, so $f \in L^p(\mathbf{R}^n)$ implies $\tilde{f} \in L^1(\mathbf{R}^n)$. To deduce (ii), we may apply B.1 once we know that $h \in C_c^\infty(\mathbf{R}^n)$ and $f \in L^1(\mathbf{R}^n)$ implies $h * f$ is continuous. However, since h is uniformly continuous, if $x_n \to x$ we have $h(x_n - t) \to h(x - t)$ uniformly in t which clearly implies continuity of $h * f$.

(iii) follows from (ii) and the fact that $\delta * f = f * \delta$.

(iv) follows from the definitions and Fubini's theorem.

Now let $\delta \in C_c^\infty(\mathbf{R}^n)$ with $\delta \geq 0$, $\delta(x) = \delta(-x)$, $\int \delta = 1$ and $\delta(x) = 0$ if $\|x\| \geq 1$. For each $\varepsilon > 0$, let $\delta_\varepsilon(x) = \varepsilon^{-n}\delta(x/\varepsilon)$. Then $\delta_\varepsilon \geq 0$, $\int \delta_\varepsilon = 1$ and δ_ε is supported on the ball around 0 of radius ε.

DEFINITION B.4. *The functions δ_ε are called an approximate identity for $\mathbf{R}^n$.*

This terminology is motivated by:

PROPOSITION B.5. *For any $1 \le p < \infty$ and $f \in L^p(\mathbf{R}^n)$, we have $\delta_\varepsilon * f \to f$ in $L^p(\mathbf{R}^n)$ as $\varepsilon \to 0$.*

PROOF: First suppose $f \in C_c(\mathbf{R}^n)$. Then the discussion preceding Definition B.2 shows $\delta_\varepsilon * f \to f$ pointwise. Furthermore $\|\delta_\varepsilon * f\|_\infty \le \|f\|_\infty$ and since f has compact support, if $\varepsilon < 1$ all $\delta_\varepsilon * f$ are supported on a common compact set. Therefore the dominated convergence theorem implies $\delta_\varepsilon * f \to f$ in $L^p(\mathbf{R}^n)$. Given any $f \in L^p$, and any $r > 0$, choose $\varphi \in C_c(\mathbf{R}^n)$ such that $\|\varphi - f\|_p < r/4$. Then in L^p,

$$\begin{aligned}
\|\delta_\varepsilon * f - f\| &\le \|\delta_\varepsilon * f - \delta_\varepsilon * \varphi\| + \|\delta_\varepsilon * \varphi - \varphi\| + \|\varphi - f\| \\
&\le 2\|\varphi - f\| + \|\delta_\varepsilon * \varphi - \varphi\| \qquad \text{(by B.3.i)} \\
&\le r/2 + \|\delta_\varepsilon * \varphi - \varphi\|.
\end{aligned}$$

For ε sufficiently small, we have $\|\delta_\varepsilon * \varphi - \varphi\| \le r/2$, and hence $\|\delta_\varepsilon f - f\| < r$.

COROLLARY B.6. $C_c^\infty(\Omega)$ *is dense in* $L^p(\Omega)$ *for any open* $\Omega \subset \mathbf{R}^n (1 \le p < \infty)$.

PROOF: It suffices to see that the closure of $C_c^\infty(\Omega)$ contains $C_c(\Omega)$. If $f \in C_c(\Omega)$, let $d = \operatorname{dist}(\operatorname{supp}(f), \partial\Omega)$ (and let $d = \infty$ if $\Omega = \mathbf{R}^n$.) If $\varepsilon < d/2$, we have $\delta_\varepsilon * f \in C_c(\Omega)$ since δ_ε is supported on the ε-ball around 0, and $\delta_\varepsilon * f \in C^\infty(\mathbf{R}^n)$ by B.3. Finally B.5 implies $\delta_\varepsilon * f \to f$ in $L^p(\Omega)$.

TOPOLOGICAL VECTOR SPACES
AND OPERATORS

1.1 Examples of spaces

A basic point of a norm on a space of functions is of course that it gives us a framework for discussing the convergence of a sequence of functions to another, say $f_n \to f$. Roughly speaking, in chapter 0 we discussed norms which enabled us to deal with uniform convergence (e.g., the norm on $BC(X)$) or L^p-convergence. In various situations, we may be interested in other types of convergence. For example we may wish to consider pointwise convergence. One can show, for example, that there is no norm on $C([0, 1])$ such that $f_n \to f$ pointwise if and only if $\|f_n - f\| \to 0$. Similarly, for $C(\mathbf{R})$, it is natural to consider $f_n \to f$ uniformly on compact subsets of $\mathbf{R}$. Once again, there is no norm on $C(\mathbf{R})$ for which this is equivalent to convergence in norm. In a somewhat different direction, instead of uniform convergence, we may wish to control the derivatives of functions as well. Thus, for smooth functions on $\mathbf{R}$, we may wish to consider $f_n \to f$ if $\frac{d^r f_n}{dx^r} \to \frac{d^r f}{dx^r}$ for all r. Here again, the framework of normed spaces is not adequate. However all these examples can be dealt with by considering a topology defined by a suitable family of seminorms.

DEFINITION 1.1.1. *A topological vector space (hereafter abbreviated by TVS) is a vector space E together with a Hausdorff topology for which the vector space operations are continuous.*

Thus, a normed space is a TVS in a natural way.

DEFINITION 1.1.2. *Let E be a vector space and $\{\| \ \|_\alpha \mid \alpha \in I\}$ a family of seminorms on E. The family is called sufficient if for all $x \in E, x \neq 0$, there is some $\alpha \in I$ such that $\|x\|_\alpha \neq 0$.*

The following is completely straightforward.

PROPOSITION 1.1.3. *If* $\{\| \ \|_\alpha\}$ *is a sufficient family of seminorms on* E, *let* E *have the topology generated by all the open* $\| \ \|_\alpha$-*balls. Then* E *is a TVS. Furthermore, for a net* $\{f_\beta\}$ *in* E *we have* $f_\beta \to f$ *in this topology if and only if for all* α *we have* $\|f_\beta - f\|_\alpha \to 0$.

The sufficiency of the family is needed in Proposition 1.1.3 to show that the topology is Hausdorff. If $\{\| \ \|_\alpha\}$ is a countable sufficient family, then the topology on E is first countable, i.e. has for each point a countable basis for the open sets containing that point. In that case, we can understand convergence by speaking of sequences rather than nets. In fact, in this case E will be metrizable.

PROPOSITION 1.1.4. *Suppose* $\{\| \ \|_n \mid 1 \le n < \infty\}$ *is a countable sufficient family of seminorms on* E. *Then the topology on* E *is metrizable.*

PROOF: For each n, let $d_n(x,y) = \frac{\|x-y\|_n}{1+\|x-y\|_n}$. Define $d(x,y) = \sum_{n=1}^{\infty} d_n(x,y)/2^n$. Then one easily verifies that d is a metric and $d(x_j, x) \to 0$ if and only if for each $n, d_n(x_j, x) \to 0$, which in turn is true if and only if $\|x_j - x\|_n \to 0$. Thus, convergence in the metric is equivalent to convergence in the topology.

EXAMPLE 1.1.5: Let X be a Hausdorff space, and for each $x \in X$ define $\| \ \|_x$ on $C(X)$ by $\|f\|_x = |f(x)|$. Then $\{\| \ \|_x \mid x \in X\}$ is a sufficient family and $f_n \to f$ is equivalent to pointwise convergence.

EXAMPLE 1.1.6: Let X be a Hausdorff space and for each compact subset $K \subset X$, define $\| \ \|_K$ on $C(X)$ by $\|f\|_K = \sup\{|f(x)| \mid x \in K\}$. Then $\{\| \ \|_K \mid K \subset X \text{ compact}\}$ is a sufficient family and $f_n \to f$ means uniform convergence on compact subsets. If X is σ-compact, i.e., $X = \bigcup_{n=1}^{\infty} U_n$ where U_n is open and $K_n = \overline{U}_n$ is compact, then the same topology is defined by the countable sufficient family $\{\| \ \|_n = \| \ \|_{K_n}\}$.

One can make similar constructions for L^p-norms. Namely, let X be a separable metric space and μ a measure on X which is finite on compact sets. Fix $p, 1 \le p \le \infty$. We say that f is locally L^p if for every compact subset $K \subset X$, $f|K \in L^p(K)$. We let $L^p_{\text{loc}}(X)$ be the space of locally L^p-functions.

EXAMPLE 1.1.7: For $K \subset X$ compact and $f \in L^p_{loc}(X)$, define $\|f\|_K = (\int_K |f|^p)^{1/p}$. Then $\{\| \ \|_K\}$ is a sufficient family and $f_n \to f$ means L^p- convergence on compact sets. As in 1.1.6, if X is σ-compact, the topology is defined by a countable sufficient family.

We now discuss examples in which the topology takes into account the size of derivatives. We first establish some notation. Let $\Omega \subset \mathbf{R}^n$ be an open set. For $i = 1, \ldots, n$, let α_i be a non-negative integer, and write $\alpha = (\alpha_1, \ldots, \alpha_n)$. For a function f on Ω we let

$$D^\alpha f = \frac{\partial^\alpha f}{\partial x_1^{\alpha_1} \cdots \partial x_n^{\alpha_n}}$$

if this exists. We set $|\alpha| = \sum_{i=1}^n \alpha_i$. We let $C^r(\Omega) = \{f \colon \Omega \to k \mid D^\alpha f$ exists and is continuous for all α with $|\alpha| \leq r\}$, and $C^\infty(\Omega) = \bigcap_{r \geq 1} C^r(\Omega)$. An element of $C^\infty(\Omega)$ is called a smooth function. By $C^0(\Omega)$ we simply mean $C(\Omega)$. We let $BC^r(\Omega) = \{f \in C^r(\Omega) \mid D^\alpha f$ is bounded on Ω, for all $|\alpha| \leq r\}$, and $BC^\infty(\Omega) = \bigcap_r BC^r(\Omega)$.

EXAMPLE 1.1.8: For each α with $|\alpha| \leq r$, define $\| \ \|_\alpha$ on $BC^r(\Omega)$ by $\|f\|_\alpha = \sup\{|D^\alpha f(x)| \mid x \in \Omega\}$. Then $\{\| \ \|_\alpha \mid |\alpha| \leq r\}$ is a (finite) sufficient family of seminorms. We have $f_n \to f$ if and only if $D^\alpha f_n \to D^\alpha f$ uniformly on Ω for all $|\alpha| \leq r$.

REMARK 1.1.9: Topologies given by a finite sufficient family of semi-norms are actually given by norms, although the way the seminorms are combined to yield a norm is not canonical. For example, if $\{\| \ \|_i \mid i = 1, \ldots, n\}$ is a sufficient family of seminorms on E, the topology is given by the norm $\|x\|' = \sum_{i=1}^n \|x\|_i$. It is given as well by the norm $\|x\|' = (\sum_{i=1}^n \|x\|_i^2)^{1/2}$. In fact, if $N \colon \mathbf{R}^n \to \mathbf{R}$ is any norm on $\mathbf{R}^n$, then $\|x\|_N = N(\|x\|_1, \ldots, \|x\|_n)$ is a norm on E which yields the appropriate topology. In a given situation one specific choice of N may be most convenient.

EXAMPLE 1.1.10: Let $\Omega \subset \mathbf{R}^n$ be open. For each α with $|\alpha| \leq r$ and each compact set $K \subset \Omega$, define $\|f\|_{\alpha,K} = \sup\{|D^\alpha f(x)| \mid x \in K\}$. Then $\{\| \ \|_{\alpha,K} \mid |\alpha| \leq r, K \subset \Omega$ compact$\}$ is a sufficient family of seminorms on $C^r(\Omega)$. We have $f_n \to f$ if and only if for all $|\alpha| \leq r, D^\alpha f_n \to D^\alpha f$ uniformly on compact sets. Since Ω is σ-compact, the topology can be given by a countable family of seminorms.

Similarly, by taking all α, we obtain a countable sufficient family of seminorms on $C^\infty(\Omega)$ such that $f_n \to f$ if and only if for all α, $D^\alpha f_n \to D^\alpha f$ uniformly on compact sets. We shall refer to this as the "C^∞- topology" on $C^\infty(\Omega)$.

We shall next consider examples in which the size of the derivatives is measured by an L^p-norm. However, we first pause to discuss completeness.

DEFINITION 1.1.11.
(a) If E is a TVS and $\{x_n\} \subset E$ is a sequence, $\{x_n\}$ is called Cauchy if for every open neighborhood U of 0, we have $x_n - x_m \in U$ for n, m sufficiently large.
(b) E is called complete if every Cauchy sequence converges.
(c) E is called a Frechet space if it is complete and the topology is given by a countable sufficient family of semi-norms.

REMARKS 1.1.12: (a) If the topology on E is given by $\{\| \ \|_\alpha\}$, then x_n is Cauchy if and only if it is $\| \ \|_\alpha$ -Cauchy for each α. I.e., for all α and $\varepsilon > 0$, there is some N such that $n, m \geq N$ implies $\|x_n - x_m\|_\alpha < \varepsilon$.
(b) Every Banach space is a Frechet space.
(c) Every Frechet space is metrizable by Proposition 1.1.4. However, the metric constructed in the proof of 1.1.4 is not a complete metric.

PROPOSITION 1.1.13. *All the examples 1.1.5–1.1.10 are complete.*

PROOF: All the proofs follow easily from Example 0.7. We shall prove completeness of $C^\infty(\Omega)$ as an illustration, the other examples following in a similar manner. Suppose $\{f_j\}$ is Cauchy. Write $\Omega = \bigcup K_i$ where K_i is compact and $K_i = \overline{U}_i$ where $U_i \subset \Omega$ is open, and $U_i \subset U_{i+1}$. For each α and each i, f_j is $\| \ \|_{\alpha,i}$-Cauchy where we have written $\| \ \|_{\alpha,i}$ for $\| \ \|_{\alpha,K_i}$. By 0.7 (a), for each α and i, there is a continuous function f_α defined on K_i such that $f_j \to f_0$ and $D^\alpha f_j \to f_\alpha$ uniformly on K_i. Suppose we knew $f_\alpha = D^\alpha f_0$ on U_i. Then we would have $f_0 \in C^\infty(\Omega)$ and $f_j \to f_0$ in the topology of $C^\infty(\Omega)$. It therefore suffices to see $f_\alpha = D^\alpha f_0$, which follows by induction from the following fact of calculus.

LEMMA 1.1.14. *Let $\partial_i f$ denote the i-th partial derivative of f.
Suppose f_j is a sequence of C^1-functions such that $f_j \to f$ uniformly on an open set $U \subset \mathbf{R}^n$. Suppose that we also have $\partial_i f_j \to g$
uniformly on U where g is continuous. Then $\partial_i f$ exists in U and
in fact $\partial_i f = g$.*

We now turn to further examples, namely the Sobolev spaces,
in which the size of derivatives is measured by L^p-norms.

EXAMPLE 1.1.15: (Sobolev spaces) Let $\Omega \subset \mathbf{R}^n$ be open. Fix
$p, 1 \leq p < \infty$, and let $\| \ \|_p$ denote the usual L^p-norm. Fix $k \geq 0$
and for $|\alpha| \leq k$, let $\|f\|_{p,\alpha} = \|D^\alpha f\|_p$. We set

$$C^\infty(\Omega)_{p,k} = \{f \in C^\infty(\Omega) \mid \|f\|_{p,\alpha} < \infty \text{ for all } |\alpha| \leq k\}.$$

Then $\{\| \ \|_{p,\alpha} \mid |\alpha| \leq k\}$ defines a finite sufficient family of seminorms on $C^\infty(\Omega)_{p,k}$. It follows that $C^\infty(\Omega)_{p,k}$ becomes a normed
space with, for example (cf. Remark 1.1.9), the norm

$$\|f\|_{p,k} = \sum_{|\alpha| \leq k} \|f\|_{p,\alpha} = \sum_{|\alpha| \leq k} \|D^\alpha f\|_p.$$

We clearly have $C_c^\infty(\Omega) \subset C^\infty(\Omega)_{p,k}$ for any p, k. Furthermore
$C^\infty(\Omega)_{p,0} = C^\infty(\Omega) \cap L^p(\Omega)$. In particular, $C^\infty(\Omega)_{p,0}$ is not complete, and in this is easily seen to be true in general. We define
$L^{p,k}(\Omega)$ to be the completion of $C^\infty(\Omega)_{p,k}$; $L^{p,k}(\Omega)$ is called the
(p,k)-Sobolev space of Ω. By Proposition A.11 we have $L^{p,0}(\Omega) = L^p(\Omega)$. While we have $C^\infty(\Omega)_{p,k} \subset L^p(\Omega)$, it is not immediately
clear that elements of $L^{p,k}(\Omega)$ can be thought of as functions (or at
least functions modulo null sets). We shall now show that they can
be, while at the same time present an alternative view of $L^{p,k}(\Omega)$.
The basic idea is the notion of a weak derivative.

We begin by recalling the following consequence of the product
rule for differentiation.

LEMMA 1.1.16. (Integration by parts). *Suppose $f \in C^\infty(\Omega)$ and
$\varphi \in C_c^\infty(\Omega)$ where $\Omega \subset \mathbf{R}^n$ is open. Then for any i,*

$$\int_\Omega f(\partial_i \varphi) = -\int_\Omega (\partial_i f)\varphi.$$

Hence, for any α

$$\int_\Omega f \cdot D^\alpha \varphi = (-1)^{|\alpha|} \int_\Omega (D^\alpha f) \cdot \varphi.$$

For convenience, when our domain of integration, say Ω, is understood, we shall write $(f, \varphi) = \int f \overline{\varphi}$ whenever $f\varphi$ is integrable.

DEFINITION 1.1.17. *Suppose* f, h *are locally integrable functions on* Ω *(i.e.* $f, h \in L^1_{loc}(\Omega)$*). We say that* h *is the weak* α*-th partial derivative of* f *on* Ω *and write* $h = D^\alpha_w f$*, if for all* $\varphi \in C^\infty_c(\Omega)$ *we have*

$$(f, D^\alpha \varphi) = (-1)^{|\alpha|}(h, \varphi).$$

LEMMA 1.1.18. *If a weak* α*-th derivative exists, it is unique (up to sets of measure 0). In particular, if* $f \in C^\infty(\Omega), D^\alpha f = D^\alpha_w f$ *for any* α*.*

PROOF: It clearly suffices to see that $h \in L^1_{loc}(\Omega)$ and $(h, \varphi) = 0$ for all $\varphi \in C^\infty_c(\Omega)$ implies $h = 0$. It suffices to see, for every $U \subset \Omega$ open with $\overline{U} \subset \Omega$ and $\overline{U}$ compact, that $h|U = 0$. Thus, by passing to $h|U$ for each such U, we need only consider the case $h \in L^1(\Omega)$ and $\overline{\Omega}$ compact. We remark that if we had $h \in L^p(\Omega)$ for any $p > 1$, we would have $h = 0$, by virtue of the isomorphism $L^p(\Omega)^* = L^q(\Omega)$, and the fact that $C^\infty_c(\Omega)$ is dense in $L^q(\Omega)$ for $q < \infty$ (B.6). We shall deduce the same for $h \in L^1(\Omega)$ by reducing to such a case. Namely, let $V \subset \Omega$ be open with $\overline{V} \subset \Omega$. Let δ_ε be an approximate identity for $\mathbf{R}^n$ (B.4). Then for ε sufficiently small we have $\delta_\varepsilon * \varphi \in C^\infty_c(\Omega)$ for any $\varphi \in C^\infty_c(V)$. Thus, $(h, \delta_\varepsilon * \varphi) = 0$. Hence $(\delta_\varepsilon * h, \varphi) = 0$ (by B.3 (iv)). However $\delta_\varepsilon * h$ is smooth and hence $(\delta_\varepsilon * h)|V \subset L^2(V)$. Since $C^\infty_c(V)$ is dense in $L^2(V)$, we have $(\delta_\varepsilon * h)|V = 0$. As $\varepsilon \to 0, \delta_\varepsilon * h \to h$ in $L^1(\mathbf{R}^n)$ (B.5), and hence $h|V = 0$. Since this is true for all such $V \subset \Omega, h = 0$.

EXAMPLE 1.1.19: Let $f \in C(\mathbf{R})$. Suppose there are finitely many points $t_1, \ldots, t_n \in \mathbf{R}, t_i < t_{i+1}$, such that $f|(t_i, t_{i+1})$ is differentiable. Then the weak derivative $d^w f$ exists and is equal to $f'(t)$ for any $t \notin \{t_1, \ldots, t_n\}$. This is a simple exercise in integration by parts. (Exercise 1.23.)

DEFINITION 1.1.20. *Let* $W^{p,k}(\Omega) = \{f \mid D_w^\alpha f$ *exists for all* $|\alpha| \leq k$ *and* $D_w^\alpha f \in L^p(\Omega)\}$. *We define a norm on* $W^{p,k}$ *by*

$$\|f\|_{p,k} = \sum_{|\alpha| \leq k} \|D_w^\alpha f\|_p.$$

(As in Remark 1.1.9, we could take $\|f\|_{p,k} = (\sum_{|\alpha| \leq k} \|D_w^\alpha f\|_p^p)^{1/p}$, etc., as may be convenient.)

We clearly have $C^\infty(\Omega)_{p,k} \subset W^{p,k}(\Omega)$ in a norm preserving way.

PROPOSITION 1.1.21. $W^{p,k}(\Omega)$ *is a Banach space. Hence* $L^{p,k}(\Omega)$ *can be realized as the closure of* $C^\infty(\Omega)_{p,k}$ *in* $W^{p,k}(\Omega)$.

PROOF: $\{f_n\}$ be a $\|\ \|_{p,k}$-Cauchy sequence in $W^{p,k}(\Omega)$. Then $D_w^\alpha f_n$ is Cauchy in $L^p(\Omega)$ for every $|\alpha| \leq k$. By completeness of $L^p(\Omega)$, we have $D_w^\alpha f_n \to f_\alpha \in L^p(\Omega)$ for some f_α. It suffices to show, letting $f_0 = f$, that $D_w^\alpha f = f_\alpha$, for then $f_n \to f$ in $W^{p,k}(\Omega)$. By definition, for all $|\alpha| \leq k$ and all $\varphi \in C_c^\infty(\Omega), (f_n, D^\alpha \varphi) = (-1)^{|\alpha|}(D_w^\alpha f_n, \varphi)$. Letting $n \to \infty$ we have $(f, D^\alpha \varphi) = (-1)^{|\alpha|}(f_\alpha, \varphi)$, and hence $f_\alpha = D_w^\alpha f$ as required.

In fact, one has equality $L^{p,k}(\Omega) = W^{p,k}(\Omega)$ for any open $U \subset \mathbf{R}^n$. We give the proof here for $\Omega = \mathbf{R}^n$. The general case is more delicate.

We first collect some simple properties of weak derivatives.

PROPOSITION 1.1.22.

(a) If $f \in W^{p,k}(\Omega)$ and $|\alpha| + |\beta| \leq k$, then $D_w^\alpha(D_w^\beta f) = D_w^{\alpha+\beta} f$.

(b) If $f \in W^{p,k}(\Omega)$ and $\varphi \in C_c^\infty(\Omega)$, then $\varphi f \in W^{p,k}(\Omega)$

(c) If $\delta \in C_c^\infty(\mathbf{R}^n)$ and $f \in W^{p,k}(\mathbf{R}^n)$, then $\delta * f \in C^\infty(\mathbf{R}^n) \cap W^{p,k}(\mathbf{R}^n)$, and $D^\alpha(\delta * f) = \delta * D_w^\alpha f$ for all $|\alpha| \leq k$.

PROOF: (a) is a straightforward exercise in the definition of weak derivative. To see (b), let $\psi \in C_c^\infty(\Omega)$. Then $\varphi\overline{\psi} \in C_c^\infty(\Omega)$. Hence

$$
\begin{aligned}
(f\varphi, \partial_j\psi) &= (f, \overline{\varphi}\partial_j\psi) \\
&= (f, \partial_j(\overline{\varphi}\psi) - \psi\partial_j\overline{\varphi}) \\
&= (f, \partial_j(\overline{\varphi}\psi)) - (f, \psi\partial_j\overline{\varphi}) \\
&= -(\partial_j^w f, \overline{\varphi}\psi) - (f\partial_j\varphi, \psi) \\
&= -((\partial_j^w f)\varphi + f\partial_j\varphi, \psi)
\end{aligned}
$$

where $\partial_j^w f$ is the weak derivative in Ω. It follows that $\partial_j^w(f\varphi) = (\partial_j^w f)\varphi + f\partial_j\varphi$. One now completes the proof by induction.

c) We have $\delta * f \in C^\infty(\mathbf{R}^n)$ by B.3 (ii), and $\delta * D_w^\alpha f \in L^p(\mathbf{R}^n)$ for $|\alpha| \leq k$ by B.3.(i). Thus, it suffices to see $D^\alpha(\delta * f) = \delta * D_w^\alpha f$. Let $\varphi \in C_c^\infty(\mathbf{R}^n)$. Then

$$
\begin{aligned}
(\delta * D_w^\alpha f, \varphi) &= (D_w^\alpha f, \delta^* * \varphi) &&\text{(B.3 (iv))} \\
&= (-1)^{|\alpha|}(f, D^\alpha(\delta^* * \varphi)) &&\text{(since } \delta^* * \varphi \in C_c^\infty(\mathbf{R}^n)) \\
&= (-1)^{|\alpha|}(f, \delta^* * D^\alpha\varphi) &&\text{(B.3 (iii))} \\
&= (-1)^{|\alpha|}(\delta * f, D^\alpha\varphi).
\end{aligned}
$$

This verifies the assertion.

PROPOSITION 1.1.23. $L^{p,k}(\mathbf{R}^n) = W^{p,k}(\mathbf{R}^n)$.

PROOF: Let $f \in W^{p,k}(\mathbf{R}^n)$. Let δ_ε be an approximate identity (B.4). Since $\delta_\varepsilon * f \in L^{p,k}(\mathbf{R}^n)$ by 1.1.22 (c), it suffices to see $\delta_\varepsilon * f \to f$ in $W^{p,k}(\mathbf{R}^n)$, as $\varepsilon \to 0$. However, for each $|\alpha| \leq k$, $\delta_\varepsilon * D_w^\alpha f \to D_w^\alpha f$ in $L^p(\mathbf{R}^n)$ by B.5 and hence by 1.1.22 (c), $D^\alpha(\delta_\varepsilon * f) \to D_w^\alpha f$ in $L^p(\mathbf{R}^n)$. This clearly implies $\delta_\varepsilon * f \to f$ in $W^{p,k}(\mathbf{R}^n)$.

In concluding this section on some basic examples of topological vector spaces, we discuss some features of dual spaces. If E is a TVS, we let $E^* = \{\lambda : E \to k \mid \lambda \text{ is linear and continuous}\}$.

LEMMA 1.1.24. (Hahn-Banach) *If the topology on E is defined by a sufficient family of seminorms, then for each $x \in E$, $x \neq 0$, there is some $\lambda \in E^*$, such that $\lambda(x) \neq 0$.*

PROOF: Let $\| \; \|$ be a seminorm in the sufficient family with $\|x\| \neq 0$. By Remark A.23 there is a linear map $\lambda : E \to k$ such that

$\lambda(x) \neq 0$ and for all $y \in E, |\lambda(y)| \leq \|y\|$. If $y_\alpha \to 0$ in E, then $\|y_\alpha\| \to 0$ and hence $|\lambda(y_\alpha)| \to 0$, which implies continuity.

DEFINITION 1.1.25. *Let E be a TVS defined by a sufficient family of seminorms.*

(a) *The weak topology on E is the topology defined by the (sufficient by 1.1.24) family of seminorms $\{\| \ \|_\lambda \mid \lambda \in E^*\}$ where $\|x\|_\lambda = |\lambda(x)|$.*

(b) *The weak-*-topology on E^* is the topology on E^* defined by the (sufficient by definition) family of seminorms $\{\| \ \|_x \mid x \in E\}$ where $\|\lambda\|_x = |\lambda(x)|$.*

Thus, $x_\alpha \to x$ in the weak topology on E if and only if for every $\lambda \in E^*, \lambda(x_\alpha) \to \lambda(x)$. Similarly, $\lambda_\alpha \to \lambda$ in the weak-*-topology on E^* if and only if $\lambda_\alpha(x) \to \lambda(x)$ for all $x \in E$.

EXAMPLE 1.1.26: Suppose $1 < p < \infty$, and $f_n \in L^p(\mathbf{R})$ with $\|f_n\|_p = 1$ and supp$(f_n) \subset [n, n+1]$. Then $f_n \to 0$ in the weak topology of $L^p(\mathbf{R})$. To see this, it suffices to see that for any $h \in L^q(\mathbf{R})$, where $p^{-1} + q^{-1} = 1$, we have $\int f_n h \to 0$ as $n \to \infty$. However, given any $\varepsilon > 0$, for n sufficiently large we have

$$\left(\int_{\mathbf{R}} |h|^q\right)^{1/q} - \left(\int_{(-\infty,n)} |h|^q\right)^{1/q} < \varepsilon.$$

Since $f_n = 0$ on $(-\infty, n)$ and $\|f_n\|_p = 1$, we have $\int |f_n h| < \varepsilon$ for n large, as required.

EXAMPLE 1.1.27: Let $E = C([0,1])$. We recall that $M([0,1])$ denotes the space of probability measures on $[0,1]$, so that we have $M([0,1]) \subset C([0,1])^*$. (Cf. A.19.) Let $\mu_0 \in M([0,1])$ be the measure supported on $\{0\}$. Let $\mu_n \in M([0,1])$ be the normalized Lebesgue measure supported on $(0, 1/n)$. For each n, we can clearly find $f_n \in C([0,1])$ such that $\|f_n\| = 1, \int f_n d\mu_n = 1/2$, and $f_n(0) = 0$. The last condition is of course the same as $\int f_n d\mu_0 = 0$. It follows that in the norm on $C([0,1])^*$, we have $\|\mu_n - \mu_0\| \geq 1/2$. On the other hand, it is clear that in some sense we should have $\mu_n \to \mu_0$. This sense is captured by the weak-*-topology. Namely, we claim that in fact $\mu_n \to \mu_0$ in the weak-*-

topology. To see this, we need to see that for each $f \in C([0,1])$ we
have $\int f d\mu_n \rightarrow \int f d\mu_0 = f(0)$. However, by continuity of f at 0,
for any $\varepsilon > 0$ we have for n sufficiently large that $|f(x) - f(0)| < \varepsilon$
on $[0, 1/n]$, and hence $|\int f d\mu_n - f(0)| < \varepsilon$.

One of the most useful features of the weak-*-topology is the
following result.

THEOREM 1.1.28. *Let E be a normed linear space. Then E_1^* (the
unit ball in E^*) is compact with the weak-*-topology.*

PROOF: The proof of this will follow easily from Tychonoff's theorem: the product of compact spaces is compact. Namely, for each
$x \in E$, let $B_x = \{c \in k \mid |c| \leq \|x\|\}$. Then B_x is compact, and
hence so is $\Omega = \Pi_{x \in E} B_x$, with the product topology. There is a
natural map $i \colon E_1^* \rightarrow \Omega$, namely $(i(\lambda))_x = \lambda(x)$. That $\lambda(x) \in B_x$
follows from the fact that $\lambda \in E_1^*$, i.e. $\|\lambda\| \leq 1$. Since a net $\omega_\alpha \in \Omega$
converges to $\omega \in \Omega$ if and only if $(\omega_\alpha)_x \rightarrow \omega_x$ for all x, it follows
that i is a homeomorphism of E_1^* with $i(E_1^*)$. It therefore suf-
fices to see that $i(E_1^*) \subset \Omega$ is a closed set. For each $x, y \in E$,
let $\Omega_{x,y} = \{\omega \in \Omega \mid \omega_{x+y} = \omega_x + \omega_y\}$, and for each $c \in k$ and
$x \in E, \Omega^{c,x} = \{\omega \in \Omega \mid \omega_{cx} = c\omega_x\}$. Then each $\Omega_{x,y}$ and $\Omega^{c,x}$
is closed in Ω and $i(E_1^*) = \bigcap_{x,y} \Omega_{x,y} \cap \bigcap_{c,x} \Omega^{c,x}$, completing the
proof.

COROLLARY 1.1.29. *Let X be a compact metric space. Then
$M(X)$ is compact with the weak-*-topology.*

PROOF: By 0.19, $M(X) = \{\lambda \in C(X)_1^* \mid \lambda(f) \geq 0 \text{ for } f \geq 0 \text{ and}$
$\lambda(1) = 1\}$. For each $f, \{\lambda \mid \lambda(f) \geq 0\}$ is closed in the weak-*-
topology. Thus, $M(X) = \bigcap_{f \geq 0} \{\lambda \mid \lambda(f) \geq 0\} \bigcap \{\lambda \mid \lambda(1) = 1\}$, and
hence is closed in $C(X)_1^*$.

REMARK: Although E^* is not in general metrizable with the weak-
*- topology even if E is separable, E_1^* will be metrizable. (See
exercise 1.10). Thus, $M(X)$ is actually a compact metrizable space
if X is.

Here is another useful fact about the weak-*-topology.

PROPOSITION 1.1.30. *Let E be a TVS defined by a sufficient family of seminorms. Then any element of $(E^*, \text{weak-}*\text{-topology})^*$ is of the form $\lambda \mapsto \lambda(x)$ for some $x \in E$.*

PROOF: Suppose $\varphi \colon E^* \to k$ is weak-* continuous and linear. Then $\varphi^{-1}(\{t \in k \mid |t| < 1\})$ is open in E^*. Hence, there are finitely many elements $x_1, \ldots, x_n \in E$ and $r_1, \ldots, r_n > 0$ such that

$$\varphi^{-1}(\{|t| < 1\}) \supset \{\lambda \in E^* \mid |\lambda(x_i)| < r_i \quad \text{for all } 1 \leq i \leq n\}.$$

In particular, if $\lambda \in E^*$ with $\lambda(x_i) = 0$ for all i, then $c\lambda(x_i) = 0$ for all i and c, and hence $|\varphi(c\lambda)| < 1$ for $c \in k$. This clearly implies $\varphi(\lambda) = 0$. We rephrase this as follows. Let $f \colon E^* \to k^n$ be given by $f(\lambda) = (\lambda(x_1), \ldots, \lambda(x_n))$. Then the above remarks simply assert that $f(\lambda) = 0$ implies $\varphi(\lambda) = 0$. It follows that φ factors through $f(E^*)$, i.e. there is a linear map $\tilde{\varphi} \colon f(E^*) \to k$ such that $\tilde{\varphi} \circ f = \varphi$. Since $f(E^*) \subset k^n$, we have for some $c_i \in k$ that $\tilde{\varphi}(a_1, \ldots, a_n) = \sum c_i a_i$ for any $(a_1, \ldots, a_n) \in f(E^*)$. Thus,

$$\varphi(\lambda) = \tilde{\varphi} f(\lambda) = \sum c_i \lambda(x_i) = \lambda(\sum c_i x_i).$$

Thus $x = \sum c_i x_i$ is the required element.

REMARK 1.1.31:
(a) In addition to $L^2(X)$, $W^{2,k}(\Omega)$ (and hence $L^{2,k}(\Omega)$) are all Hilbert spaces. Namely, for $f, h \in W^{2,k}(\Omega)$ we define $\langle f, h \rangle_k = \sum_{|\alpha| \leq k} \langle D^\alpha f, D^\alpha h \rangle_0$ where $\langle \ , \ \rangle_0$ denotes the ordinary inner product on $L^2(\Omega)$. Thus for this inner product we have $\|f\|_{2,k} = (\sum \|D^\alpha f\|_{2,0}^2)^{1/2}$.
(b) If E is a Hilbert space, then we have a natural identification of E with E^*. The weak-*-topology on E^* thus defines a topology on E, and this is clearly just the weak topology on E. Thus, the unit ball in E is compact with the weak topology.

1.2. Examples of operators

We begin by establishing some notation, generalizing that of A.13.

If E, F are TVS's, we let $B(E, F)$ be the space of continuous linear maps $E \to F$. If $E = F$, we write $B(E, F) = B(E)$. If $T \in B(E, F)$ is bijective with a continuous inverse, then T will be called an isomorphism of E and F. If $E = F$, then T will be called an automorphism of E and we let $\mathrm{Aut}(E) \subset B(E)$ be the set of automorphisms. If E is a normed space, we let $\mathrm{Iso}(E) \subset \mathrm{Aut}(E)$ be the set of isometric automorphisms. If $\dim E < \infty$, we shall often write $\mathrm{Aut}(E) = \mathrm{GL}(E) \cong \mathrm{GL}(n, k)$ where $n = \dim E$.

Our first examples are multiplication operations.

EXAMPLE 1.2.1: Let (X, μ) be a measure space and $\varphi \in L^\infty(X)$. Define $M_\varphi : L^p(X) \to L^p(X)$ by $M_\varphi f = \varphi \cdot f$. Then for any $1 \leq p \leq \infty$, M_φ is bounded, and in fact $\|M_\varphi\| = \|\varphi\|_\infty$. To see this, we clearly have $\|\varphi f\|_p \leq \|\varphi\|_\infty \|f\|_p$, so that $\|M_\varphi\| \leq \|\varphi\|_\infty$. On the other hand, choose a measurable A with $0 < \mu(A) < \infty$ and $|\varphi(x)| \geq \|\varphi\|_\infty - \varepsilon$ for $x \in A$. Then the characteristic function $\chi_A \in L^p(X)$ and $\|\varphi \chi_A\|_p \geq (\|\varphi\|_\infty - \varepsilon)\|\chi_A\|_p$. This establishes the asserted equality.

EXAMPLE 1.2.2: For $\Omega \subset \mathbf{R}^n$ open and $\varphi \in BC^k(\Omega)$, define $M_\varphi : C^\infty(\Omega)_{p,k} \to C^\infty(\Omega)_{p,k}$ by $M_\varphi f = \varphi f$. One easily checks that this is bounded (and in fact $\|M_\varphi\| \leq c_k \|\varphi\|_{BC^k(\Omega)}$ for some constant c_k), and hence extends to a bounded operator $M_\varphi : L^{p,k}(\Omega) \to L^{p,k}(\Omega)$.

REMARK 1.2.3: We can interpret the above as follows. Let $B(L^p(X))$ and $B(L^{p,k}(\Omega))$ have the norm topologies (0.15). Then the map M (i.e. $\varphi \mapsto M_\varphi$) defines an isometry $M : L^\infty(X) \to B(L^p(X))$ and a bounded injection $M : BC^k(\Omega) \to B(L^{p,k}(\Omega))$.

We now turn to translation operators.

If X is a set and $\varphi : X \to X$ is a bijection, then φ defines a "translation operator" T_φ on spaces of functions on X. Namely, if $f : X \to k$, then we have $(T_\varphi f)(x) = f(\varphi^{-1}(x))$, i.e. $T_\varphi f = f \circ \varphi^{-1}$. With various hypotheses on X we can form various function spaces, and T_φ will under suitable hypotheses on φ define a continuous operator on this space.

EXAMPLE 1.2.4: (a) Let X be a Hausdorff space and $\varphi \in \mathrm{Homeo}(X)$ (where $\mathrm{Homeo}(X)$ is the group of homeomorphisms of X.) Then $T_\varphi : BC(X) \to BC(X)$ is an isometry, with

$(T_\varphi)^{-1} = T_{\varphi^{-1}}$.

(b) Suppose X is a locally compact separable metric space and $\varphi \in \text{Homeo}(X)$. Give $C(X)$ the structure of a Frechet space with the topology of uniform convergence on compact sets. Then $T_\varphi \colon C(X) \to C(X)$ is an isomorphism (with inverse T_φ^{-1}).

We now consider translations on L^p-spaces. We need a preliminary definition.

DEFINITION 1.2.5. *If (X, μ) is a measure space and $\varphi \colon X \to X$ is a measurable map, we say that μ is φ-invariant, or that φ is measure preserving (μ being understood) if $\varphi_* \mu = \mu$ where we define $(\varphi_* \mu)(A) = \mu(\varphi^{-1}(A)))$ for all measurable $A \subset X$. (This is equivalent to $\int (f \circ \varphi)\, d\mu = \int f d\mu$ for all bounded Borel functions f. If φ is bijective with measurable inverse, then μ is φ-invariant if and only if it is φ^{-1}- invariant.)*

EXAMPLE 1.2.6: (a) For $X = \mathbf{R}^n$ with Lebesgue measure, and any $t \in \mathbf{R}^n, \varphi(x) = x + t$ is measure preserving.

(b) For $X = \mathbf{R}^n$ with Lebesgue measure and any $\varphi \in \text{GL}(n, \mathbf{R})$, (i.e. φ is a linear automorphism of $\mathbf{R}^n$), then $\varphi \colon \mathbf{R}^n \to \mathbf{R}^n$ is measure preserving if and only if $|\det \varphi| = 1$. This is essentially the definition of the determinant.

DEFINITION 1.2.7. *Let $\Omega \subset \mathbf{R}^n$ be open. A map $\varphi \colon \Omega \to \Omega$ is called a diffeomorphism if φ is a homeomorphism and φ and φ^{-1} are smooth.*

We remark that by the chain rule this implies $d\varphi_x$ is invertible for all $x \in \Omega$.

EXAMPLE 1.2.8: If $\varphi \colon \Omega \to \Omega$ is a diffeomorphism, then φ preserves Lebesgue measure if and only if $|\det d\varphi_x| = 1$ for all $x \in \Omega$. To see this, we simply recall that the change of variable formula for integration implies that for any bounded Borel function

$$\int (f \circ \varphi)|\det d\varphi| d\mu = \int f d\mu.$$

Thus

$$\int (f \circ \varphi) d\mu = \int f d\mu$$

for all such f if and only if $|\det d\varphi| = 1$ a.e., and since $x \mapsto d\varphi_x$ is continuous, this is equivalent to $|\det d\varphi_x| = 1$ for all x.

EXAMPLE 1.2.9: Suppose $\varphi\colon X \to X$ is a measure preserving bijection with a measurable inverse. Then $T_\varphi\colon L^p(X) \to L^p(X)$ is an isometric isomorphism for all $p, 1 \leq p \leq \infty$. This is immediate from

$$\int |f \circ \varphi|^p d\mu = \int (|f|^p \circ \varphi)d\mu = \int |f|^p.$$

(For $p = \infty$, T_φ will be an isometric isomorphism simply with the hypotheses that $\mu(\varphi^{-1}(A)) = 0$ if and only if $\mu(A) = 0$ rather than the stronger assumption that μ be invariant. Thus, for example, if $\varphi\colon \Omega \to \Omega$ is any diffeomorphism, T_φ will be an isometric isomorphism of $L^\infty(\Omega)$.)

EXAMPLE 1.2.10: Suppose $\varphi\colon \Omega \to \Omega$ is a diffeomorphism. Then $T_\varphi\colon C^\infty(\Omega) \to C^\infty(\Omega)$ is an isomorphism where $C^\infty(\Omega)$ has the Frechet space topology defined in 1.1.10.

We now define differential operators.

DEFINITION 1.2.11. *Let $\Omega \subset \mathbf{R}^n$ be open. A differential operator of order $\leq r$ on Ω is an operator $D\colon C^\infty(\Omega) \to C^\infty(\Omega)$ of the form*

$$(Df)(x) = \sum_{|\alpha| \leq r} a_\alpha(x)(D^\alpha f)(x)$$

where $a_\alpha \in C^\infty(\Omega)$. That is,

$$D = \sum_{|\alpha| \leq r} M_{a_\alpha} \circ D^\alpha,$$

where M is multiplication.

D will be continuous if $C^\infty(\Omega)$ has the C^∞-topology (1.1.10). We say that D has order r if it has order $\leq r$ and some $a_\alpha(x) \neq 0$ where $|\alpha| = r$. A differential operator will define continuous operators on other spaces under suitable hypotheses.

EXAMPLE 1.2.12: Suppose D is of order r and all coefficients $a_\alpha \in BC^\infty(\Omega)$. Then D induces continuous operators (for $r \leq k$) $D: BC^k(\Omega) \to BC^{k-r}(\Omega)$ and $D: L^{p,k}(\Omega) \to L^{p,k-r}(\Omega)$.

EXAMPLE 1.2.13: (Embedding operators). There are numerous continuous inclusions between the various spaces we have defined. Thus, for $k \geq \ell$, we have $C^k(\Omega) \hookrightarrow C^\ell(\Omega)$ is continuous, and $L^{p,k}(\Omega) \hookrightarrow L^{p,\ell}(\Omega)$ is continuous. There are some non-obvious inclusions as well. (See Section 5.2, for example.)

EXAMPLE 1.2.14: (Integral operators). For simplicity we shall consider integral operators on $L^2(X)$. We let (X, μ) be a measure space and $K \in L^2(X \times X, \mu \times \mu)$. Define $T_K : L^2(X) \to L^2(X)$ by $(T_K f)(x) = \int_X K(x, y) f(y) d\mu(y)$. We need to verify that $T_K f \in L^2(X)$. For each $x \in X$, let $K_x(y) = K(x, y)$. Then by Fubini's theorem, $K_x \in L^2(X)$ for a.e. $x \in X$. Thus, $(T_K f)(x) = \langle K_x, \overline{f} \rangle$ and hence is defined (a.e.) To see $T_K f \in L^2(X)$, observe that

$$\begin{aligned}
\|T_K f\|_2^2 &= \int_X |\langle K_x, \overline{f} \rangle|^2 dx \\
&\leq \int_X \|K_x\|^2 \|f\|^2 dx \\
&= \|f\|^2 \int_X \|K_x\|^2 dx \\
&= \|f\|^2 \int_X \int_X |K(x, y)|^2 dx dy.
\end{aligned}$$

Thus, $\|T_K f\|_2 \leq \|f\| \, \|K\|_{L^2(X \times X)}$. This not only shows that $T_K f \in L^2(X)$, but that $\|T_K\| \leq \|K\|_{L^2(X \times X)}$.

EXAMPLE 1.2.15: In the above example, take $X = \{1, \ldots, n\}$ with counting measure, then $K = \{K(i, j)\}$ is simply an $n \times n$ matrix. (As usual, we then write $K(i, j) = K_{ij}$). The formula for T_K is just the usual correspondence of a linear operator on K^n to an $n \times n$ matrix. We can thus view the functions K on $X \times X$ as a generalization of a matrix and integral operators as a generalization of the formula for applying matrices to vectors.

REMARK 1.2.16: If X is compact, $\mu(X) < \infty$, and $K \in C(X \times X)$, then $T_K(L^2(X)) \subset C(X)$. This follows from the facts that the map $X \to C(X)$ given by $x \mapsto K_x$ is continuous (which is simply a point set topology exercise), the inclusion $C(X) \to L^2(X)$ is continuous, and $(T_K f)(x) = \langle K_x, \overline{f} \rangle$.

REMARK 1.2.17: Let E be a separable Hilbert space with orthonormal basis $\{e_i\}$. Then any bounded linear operator T is determined by $\{Te_j\}$, and hence by $\{\langle Te_j, e_i \rangle\}$. Thus, exactly as in finite dimensions, we associate to T a matrix (now infinite), namely $T_{ij} = \langle Te_j, e_i \rangle$ from which we can recover T by "matrix multiplication". I.e. if $x = \Sigma a_i e_i$, we have $(Tx)_i = \Sigma T_{ij} a_j$. Consider now the example $E = \ell^2(\mathbb{N})$ with orthonormal basis $e_i(j) = \delta_{ij}$. If $M_\varphi \in B(\ell^2(\mathbb{N}))$ is a multiplication operator for $\varphi \in \ell^\infty(\mathbb{N})$, then M_φ has a "diagonal" matrix. I.e. $(M_\varphi)_{ij} = 0$ if $i \neq j$. If $K \in \ell^2(\mathbb{N} \times \mathbb{N})$, then the matrix for T_K is just K itself, i.e. $(T_K)_{ij} = K(i,j)$.

REMARK 1.2.18: Unlike the case of finite dimensions, not every $T \in B(L^2(X))$ is of the form T_K for some $K \in L^2(X \times X)$. See 3.1.5, for example.

We now turn to adjoint operators.

DEFINITION 1.2.19. Let E, F be TVS's, and $T: E \to F$ a continuous linear map. Define the adjoint T^* of T by $T^*: F^* \to E^*, T^*(\lambda) = \lambda \circ T$.

REMARK 1.2.20: Clearly T^* is linear. Its continuity depends upon the topologies on E^*, F^*. It is immediate from the definitions that T^* is always continuous if E^*, F^* have the weak-*-topologies.

LEMMA 1.2.21. If E, F are normed, then $\|T^*\| = \|T\|$.

PROOF: For $\lambda \in F^*, x \in E$ we have $|(T^*\lambda)(x)| = |\lambda(Tx)| \leq \|\lambda\| \|T\| \|x\|$. Hence $\|T^*\lambda\| \leq \|T\| \|\lambda\|$, which implies $\|T^*\| \leq \|T\|$. For the reverse inequality, let $\varepsilon > 0$ and choose x with $\|x\| = 1$ and $\|Tx\| \geq \|T\| - \varepsilon$. By Hahn-Banach, we can find $\lambda \in F^*$ such that $\|\lambda\| = 1$ and $|\lambda(Tx)| = \|Tx\|$. Then $|(T^*\lambda)(x)| = |\lambda(Tx)| = \|Tx\| \geq \|T\| - \varepsilon$. Since $\|x\| = \|\lambda\| = 1$, this implies $\|T^*\| \geq \|T\| - \varepsilon$.

EXAMPLE 1.2.22: Suppose E is a Hilbert space. Let $i: E \to E^*$ be the bijection $(i(x))(y) = \langle y, x \rangle$. If $T \in B(E)$, then $T^* \in B(E^*)$ and hence defines an operator in $B(E)$ via the identification i

I.e., we obtain an operator $T' = i^{-1} \circ T^* \circ i$. To see what T' is we observe that $i \circ T' = T^* \circ i$ implies that for all $x, y \in E$ we have $\langle Tx, y \rangle = \langle x, T'y \rangle$. By a standard abuse of notation, we shall denote T' by T^*. Thus, for any $T \in B(E)$, we have $T^* \in B(E)$ with $\|T^*\| = \|T\|$ and T^* is characterized by the equation $\langle Tx, y \rangle = \langle x, T^*y \rangle$. If $\{e_i\}$ is an orthonormal basis, then the matrix for T^* with respect to $\{e_i\}$ is $T^*_{ij} = \langle T^*e_j, e_i \rangle = \langle e_j, Te_i \rangle = \overline{\langle Te_i, e_j \rangle} = \overline{T}_{ji}$. Thus, as in finite dimensions, T^* is the conjugate transpose.

DEFINITION 1.2.23. (a) *If E is a Hilbert space and $T \in B(E)$, then T is called self-adjoint if $T = T^*$.*
(b) *An operator $U \in B(E)$ is called unitary if U is an isometric isomorphism, or equivalently, if U is surjective and $\langle Ux, Uy \rangle = \langle x, y \rangle$ for all x, y. Thus, U is unitary if and only if $U^{-1} = U^*$.*
(c) *If E is a Hilbert space, we shall often denote $\mathrm{Iso}(E)$ by $U(E)$.*

EXAMPLE 1.2.24: If $\varphi \in L^\infty(X)$, let $M_\varphi \in B(L^2(X))$ be the corresponding multiplication operator. Then $(M_\varphi)^* = M_{\overline{\varphi}}$. This is immediate as it suffices to see for all $f, h \in L^2(X)$, that $\langle \varphi f, h \rangle = \langle f, \overline{\varphi}h \rangle$ which is clear. Thus, M_φ is self-adjoint if and only if $\varphi = \overline{\varphi}$ a.e.; that is, φ is real valued. Similarly, M_φ is unitary if and only if $M_\varphi M_{\overline{\varphi}} = I$, i.e. $M_{\varphi\overline{\varphi}} = I$. Thus, M_φ is unitary if and only if $\varphi\overline{\varphi} = 1$ a.e., that is, $|\varphi(x)| = 1$ a.e.

1.3. Operator topologies and groups of operators

Let E, F be TVS's and $B(E, F)$ the space of continuous linear maps $E \to F$. As in the case $F = k$ (so that $B(E, F)$ becomes E^*), there are various topologies of interest that we can put on $B(E, F)$. We indicate these here only for E, F normed, but it is easy to see when and how to generalize to topologies defined by a family of seminorms.

DEFINITION 1.3.1. *Let E, F be normed spaces.*
(a) *The norm topology on $B(E, F)$ is just that defined by the norm on $B(E, F)$. Thus, $T_\alpha \to T$ if and only if $\|T_\alpha - T\| \to 0$.*
(b) *For $x \in E$, define the seminorm $\| \ \|_x$ on $B(E, F)$ by $\|T\|_x = \|Tx\|$. The topology on $B(E, F)$ defined by the family $\{\| \ \|_x \mid x \in E\}$ is called the strong operator topology. Thus*

$T_\alpha \to T$ if and only if $T_\alpha x \to Tx$ in F for all $x \in E$; i.e. $\|T_\alpha x - Tx\|_F \to 0$ for all $x \in E$. When $F = k$, the strong operator topology on E^* is the same as the weak-*-topology.

(c) For $x \in E$ and $\lambda \in F^*$, define the seminorm $\| \ \|_{x,\lambda}$ on $B(E,F)$ by $\|T\|_{x,\lambda} = |\lambda(Tx)|$. The topology on $B(E,F)$ defined by the family $\{\| \ \|_{x,\lambda} \mid x \in E, \lambda \in F^*\}$ is called the weak operator topology. Thus $T_\alpha \to T$ if and only if for all $x \in E$, $T_\alpha x \to Tx$ in F where F has the weak topology, i.e. $\lambda(T_\alpha x) \to \lambda(Tx)$ for all $\lambda \in F^*$. When $B(E,F) = E^*$, the weak operator topology also coincides with the weak-*- topology.

EXAMPLE 1.3.2: Let E be a Hilbert space, $\{e_i\}$ an orthonormal basis. Suppose $\|T_n\|, \|T\| \leq 1$. Then $T_n \to T$ in the weak operator topology if and only if for all i, j the matrix entry $(T_n)_{ij} \to T_{ij}$. Convergence in the strong operator topology is equivalent to convergence of each column in the L^2-norm, i.e. convergence of $T_n e_j \to Te_j$ in L^2. Convergence in norm requires convergence of the columns uniformly as we vary the columns.

In many situations, it is natural to consider not just one operator but rather a group of operators. For example, for each $t \in \mathbf{R}^n$ we have $\varphi_t \colon \mathbf{R}^n \to \mathbf{R}^n$ defined by $\varphi_t(x) = x - t$. In turn, each φ_t defines a translation operator $T_t = T_{\varphi_t}$, $T_t \colon L^p(\mathbf{R}^n) \to L^p(\mathbf{R}^n)$. Here it is clearly natural to consider at once the family of operators $\{T_t \mid t \in \mathbf{R}^n\}$ rather than a single fixed operator. To put this example in a suitable context, we recall the notion of a topological group and a group action.

DEFINITION 1.3.3. *A topological group is a group G together with a Hausdorff topology such that the group operations are continuous.*

EXAMPLE 1.3.4:
(a) $\mathbf{R}^n$ with addition.
(b) $\mathbf{R}^\times = \mathbf{R} - \{0\}$ with multiplication.
(c) $\mathrm{GL}(n, \mathbf{R}) = n \times n$ invertible matrices. Since $\mathrm{GL}(n, \mathbf{R}) \subset \mathbf{R}^{n \times n}$ it has a natural topology and $\mathrm{GL}(n, \mathbf{R})$ is a topological group.
(d) Closed subgroups of $\mathrm{GL}(n, \mathbf{R})$ are topological groups. For example, $\mathrm{SL}(n, \mathbf{R}) = \{A \in \mathrm{GL}(n, \mathbf{R}) \mid \det(A) = 1\}$, and $O(n) = \{A \in \mathrm{GL}(n, \mathbf{R}) \mid \|Ax\| = \|x\| \text{ for all } x \in \mathbf{R}^n\}$.

(e) If E is a normed space, Iso(E) is a topological group with the strong operator topology. In particular, if E is a Hilbert space, $U(E)$ is a topological group with the strong operator topology. In fact, the strong operator topology and the weak operator topology coincide on $U(E)$. (See exercise 1.21.)

DEFINITION 1.3.5. A (left) action of a group G on a space X is a map $G \times X \rightarrow X$, which we denote by $(g, x) \mapsto g \cdot x$, such that $(gh) \cdot x = g \cdot (h \cdot x)$ and $e \cdot x = x$. If G is a topological group and X is a topological space, we say the action is continuous if $G \times X \rightarrow X$ is a continuous map. In that case, for each $g \in G$ the map $x \mapsto g \cdot x$ is a homeomorphism φ_g of X with inverse $\varphi_{g^{-1}}$.

EXAMPLES 1.3.6: (a) $\mathbf{R}^n$ acts on itself by translation. I.e. for $t \in \mathbf{R}^n$, we have $\varphi_t(x) = x + t$.
(b) $O(n)$ acts on S^{n-1} by rotation (i.e. matrix multiplication). More precisely, we have the action $O(n) \times S^{n-1} \rightarrow S^{n-1}$ given by $A \cdot x = A(x)$, ordinary matrix multiplication.
(c) Similarly, we have GL$(n, \mathbf{R})$ acting on $\mathbf{R}^n$.

DEFINITION 1.3.7. If G is a group and V is a vector space, a representation of G on V is a homomorphism $G \rightarrow GL(V)$, where $GL(V)$ is the group of invertible linear maps on V. If E is a TVS, a representation of G on E is a homomorphism $\pi \colon G \rightarrow Aut(E)$. If G is a topological group, we speak of a representation of G on E being a continuous representation when $Aut(E) \subset B(E)$ has a given topology (e.g. strong operator, weak operator). If E is a normed space, a representation π is called an isometric representation if $\pi(G) \subset Iso(E) \subset Aut(E)$, and if E is a Hilbert space, an isometric representation will also be called a unitary representation.

REMARK 1.3.8: (a) A continuous action of a group on X gives a group of homeomorphisms of X. Each homeomorphism yields a translation operator as in section 1.2. We thus obtain, under various assumptions on the action, various representations of G on various spaces of functions.
(b) A representation $\pi \colon G \rightarrow Aut(E)$ is continuous for the strong operator topology if and only if it is continuous at $e \in G$. (See exercise 1.12.)

As a basic example, we have:

PROPOSITION 1.3.9. *Let G be a topological group acting continuously on a locally compact space X. Let $C_c(X)$ be the space of compactly supported functions with the norm topology. Let $\pi: G \to \text{Iso}(C_c(X))$ be given by $\pi(g)f)(x) = f(g^{-1}x)$. Then π is continuous where $\text{Iso}(C_c(X))$ has the strong operator topology.*

PROOF: It suffices to see (by 1.3.8(b)) that if $f \in C_c(X)$ and $\varepsilon > 0$, then for all g in a neighborhood of $e \in G$ we have $\|\pi(g)f - f\| < \varepsilon$. Since $\text{supp}(f)$ is compact and X is locally compact, we can find an open set $U \subset X$ with compact closure such that $\text{supp}(f) \subset U$. Then for each $x \in \text{supp}(f)$, continuity of the action implies there is an open neighborhood U_x of x and an open neighborhood W_x of e in G such that $W_x \cdot U_x \subset U$. We have by compactness that for some finite set $x_1, \ldots, x_n \in \text{supp}(f)$ that $\text{supp}(f) \subset \bigcup_{i=1}^{n} U_{x_i}$. Then $W = \bigcap W_{x_i}$ will be an open neighborhood of e such that $W \cdot \text{supp}(f) \subset U$. This implies for $g \in W$, that $\text{supp}(\pi(g)f) \subset \overline{U}$. Therefore, we need only show that for g in an open neighborhood $W' \subset W$ that $|f(g^{-1}x) - f(x)| < \varepsilon$ for $x \in \overline{U}$.

We achieve this by repeating the same sort of argument. For each $x \in \overline{U}$, we can choose an open neighborhood U_x' of x such that $|f(y) - f(x)| < \varepsilon/2$ for $y \in U_x'$. By continuity of the action, we can choose open neighborhoods Z_x of e in G and Y_x such that $Z_x Y_x \subset U_x'$. Choose a (new) finite set $x_1, \ldots, x_r$ such that $\overline{U} \subset \bigcup_{i=1}^{r} Y_{x_i}$. Let $W' = W \cap \bigcap_{i=1}^{r} Z_{x_i}$. Then if $g \in (W')^{-1}$ (which is open since taking inverse in G is a homeomorphism) and $y \in \overline{U}$, we have for some i that $y \in Y_{x_i} \subset U_{x_i}$ and hence $g^{-1}y \subset U_{x_i}'$. Thus

$$|f(g^{-1}y) - f(y)| \leq |f(g^{-1}y) - f(x_i)| + |f(x_i) - f(y)| \leq \varepsilon/2 + \varepsilon/2.$$

This completes the proof.

For example, this shows that the representation $\pi: \mathbf{R}^n \to \text{Iso}(C_c(\mathbf{R}^n))$ given by $(\pi(t)f)(x) = f(x - t)$ is continuous.

We now turn to the analogous result for actions on L^p-spaces. This will imply, for example, that the representation $\pi: \mathbf{R}^n \to \text{Iso}(L^p(\mathbf{R}^n))$ given by the same formula as above is continuous (for the strong operator topology) for any $1 \leq p < \infty$.

PROPOSITION 1.3.10. *Let X be a locally compact metrizable space and G a topological group acting continuously on X. Suppose μ is a measure on X which is G- invariant, i.e. invariant for every homeomorphism φ_g ($g \in G$) of X. Suppose further that $\mu(A) < \infty$ for every compact subset $A \subset X$. Then for $1 \le p < \infty$, the representation $\pi \colon G \to \mathrm{Iso}(L^p(X))$, $(\pi(g)f)(x) = f(g^{-1}x)$, is continuous (for the strong operator topology).*

PROOF: To see that $g \to \pi(g)f$ is continuous for all $f \in L^p(X)$, it suffices (cf. exercise 1.13) to see this for $f \in C_c(X)$, since the latter is dense (A.10). But if $g_\alpha \to e$, we have already seen in Proposition 1.3.9 that $\pi(g_\alpha)f \to f$ uniformly and that we may suppose $\mathrm{supp}(\pi(g_\alpha)f), \mathrm{supp}(f) \subset \overline{U}$ for some fixed compact set $\overline{U}$. This easily implies convergence in L^p, which is what is required.

EXAMPLE 1.3.11: The representation π in Proposition 1.3.10 is not continuous in general if we give $\mathrm{Iso}(L^p(X))$ the norm topology. For example, let $\mathbf{R}$ act on $\mathbf{R}$ by translations. Then for any $t > 0$ we have

$$\|(\pi(t) - I)\chi_{[0,t]}\|_p = \|\chi_{[t,2t]} - \chi_{[0,t]}\|_p \ge \|\chi_{[0,t]}\|_p.$$

Thus, $\|\pi(t) - I\| \ge 1$ for any $t > 0$, and hence we cannot have $\pi(t) \to \pi(0) = I$ as $t \to 0$.

DEFINITION 1.3.12. *If $\pi \colon G \to GL(V)$ is a representation, we define the adjoint of π by $\pi^* \colon G \to GL(V^*)$, $\pi^*(g) = \pi(g^{-1})^*$.*

LEMMA 1.3.13. *Let π be a representation of G on a TVS E. Then π^* will be a representation of G on E^* if either:*

  (i) *E^* has the weak-*-topology;*
       or,
  (ii) *E is normed and E^* has the norm topology.*

(The issue here is whether $\pi(g)^* \colon E^* \to E^*$ is continuous. We are not discussing continuity of the representation in g.)

The proof is immediate from the definitions.

EXAMPLE 1.3.14: (a) Let G and X satisfy the hypotheses of
Proposition 1.3.10. Let π be the representation of G on $L^p(X)$, for
$1 \leq p < \infty$. Then π^* will be the representation of G on $L^q(X)$
($p^{-1} + q^{-1} = 1$) where we identify $L^p(X)^* \cong L^q(X)$.

(b) Let X be a compact metric space and suppose G acts continuously on X. Then G acts on $C(X)$, say by the representation π.
Thus G acts on $C(X)^*$ by π^* where $(\pi^*(g)(\lambda))(f) = \lambda(\pi(g)^{-1}f)$.
Since $f \geq 0$ implies $\pi(g)f \geq 0$, if $\lambda = \lambda_\mu$ for some measure μ
(cf. A.19), then $\pi^*(g)$ will also be of the form λ_ν. One can check
(exercise 1.24) that $\nu = g_* \mu$ where g is identified with the homeomorphism of X it defines by the action and $g_* \mu$ is as in Definition
1.2.5, i.e. $(g_* \mu)(A) = \mu(g^{-1}A)$.

Problems for Chapter 1

1.1. Show that any two norms on a finite dimensional vector space are equivalent.

1.2. If dim $E < \infty$, show that any topology given by a sufficient family of seminorms is given by a norm.

1.3. Show that any finite dimensional normed space is Banach.

1.4. If E is a Hilbert space show that dim$E < \infty$ if and only if there is a compact neighborhood of 0.

1.5. Show there is no norm on $C([0,1])$ such that convergence in this norm is equivalent to pointwise convergence.

1.6. If E is a TVS, a subset $U \subset E$ is called balanced if $x \in U$ and $c \in k$ with $|c| \leq 1$ implies $cx \in U$. If U is a convex (see Definition 2.1.2) balanced neighborhood of $0 \in E$, let $\|x\| = \inf\{c \in \mathbf{R} \mid c > 0, x/c \in U\}$. Show that $\|\ \|$ is a seminorm on U.

1.7. If E is a TVS, E is called locally convex (hereafter, LCTVS), if every neighborhood of $0 \in E$ contains a convex balanced open neighborhood of 0.

   (a) Show that E is locally convex if and only if the topology is defined by a sufficient family of seminorms.

   (b) If $k = \mathbf{R}$, show that E is locally convex if and only if every neighborhood of $0 \in E$ contains a convex open neighborhood of 0.

1.8. If E is a TVS, show that E is locally convex and metrizable if and only if the topology is defined by a countable sufficient family of seminorms.

1.9. Suppose E is a normed space and $F \subset E$ is a closed linear subspace. Define $N: E/F \to \mathbf{R}$ by $N(x+F) = \inf\{\|x+y\| \mid y \in F\}$.

   (a) Show that N is a norm on E/F.

   (b) Show the natural map $E \to E/F$ is continuous.

   (c) If E' is another normed space and $T \in B(E, E')$ with $T(F) = 0$, show that the induced map $E/F \to E'$ is continuous.

   (d) If E is Banach, show that E/F is as well.

1.10 If E is a separable normed space, show that the unit ball $E_1^* \subset E^*$ is a separable metrizable space with the weak-*-topology.

1.11. (a) If E is a normed space and $\mathrm{Iso}(E)$ the group of isometric isomorphisms of E, show $\mathrm{Iso}(E)$ is a topological group with the strong operator topology.

(b) Show that $\mathrm{Iso}(E)$ acts continuously on E_1^* where the latter has the weak-*-topology.

1.12. If G is a topological group and E is a normed space, show that a representation $\pi\colon G \to \mathrm{Iso}(E)$ is continuous if and only if it is continuous at $e \in G$.

1.13. Suppose E is a normed space and A_n, $A \in B(E)$, with $\|A_n\|$, $\|A\| \leq C$ for some $C \in \mathbf{R}$. Show $A_n \to A$ in the strong operator topology if and only if there is a dense set $E_0 \subset E$ such that $A_n x \to A x$ for all $x \in E$.

1.14. Let $\Omega \subset \mathbf{R}^n$ be open and $D\colon C^\infty(\Omega) \to C^\infty(\Omega)$. A differential operator $D^*\colon C^\infty(\Omega) \to C^\infty(\Omega)$ is called a formal adjoint of D if for all $\varphi \in C^\infty(\Omega)$ and $\psi \in C_c^\infty(\Omega)$, we have $(D\varphi, \psi) = (\varphi, D^*\psi)$ (where $(f, h) = \int_\Omega f\overline{h}$).

(a) If a formal adjoint exists, show it is unqiue.

(b) Show that every D has a formal adjoint. (Hint: first consider the cases $D = D^\alpha$ and $D = M_a, a \in C^\infty(\Omega)$.)

(c) Show $D^{**} = D$.

1.15. Let $E = L^\infty(\mathbf{R})$. For each n, define $\lambda_n \in E^*$ by $\lambda_n(f) = (2n)^{-1} \int_{-n}^n f$.

(a) Show $\lambda_n \in E_1^*$.

(b) Using compactness of E_1^*, let λ be a weak-* accumulation point of $\{\lambda_n\}$. Show λ is not of the form λ_f for any $f \in L^1(\mathbf{R})$ where $\lambda_f(h) = \int fh$. (Thus, $L^\infty(\mathbf{R}) \neq L^1(\mathbf{R})^*$).

1.16. If X is a topological space, let $M(X)$ be the set of probability measures on X. We have a natural inclusion $M(X) \hookrightarrow BC(X)^*$. Give an example to show that if X is not compact, $M(X)$ need not be compact in the weak-*-topology.

1.17. Let $A \in GL(n, \mathbf{R})$ and $T_A\colon C^\infty(\mathbf{R}^n) \to C^\infty(\mathbf{R}^n)$.

(a) Show that T_A induces a bounded operator on $L^{p,k}(\mathbf{R}^n)$.

(b) Show that T_A induces an isometry of $L^{2,k}(\mathbf{R}^n)$ (with respect to some equivalent norm) if and only if A is orthogonal.

1.18. Show $C_c^\infty(\mathbf{R}^n)$ is dense in $L^\infty(\mathbf{R}^n)$ with the weak-*-topology, but not the norm topology.

1.19. (a) Suppose E, F are Banach spaces and $\mathcal{A} \subset B(E, F)$. Suppose that for all $x \in E, \{\|Tx\| \mid T \in \mathcal{A}\}$ is bounded in $\mathbf{R}$.

Show $\{\|T\|\}$ is bounded. Hint: Let $A_N = \{x \in E \mid \|Tx\| \leq N$ for all $T \in \mathcal{A}\}$. Then use the Baire category theorem.

(b) Show that any sequence in E that converges in the weak topology is bounded. Show any sequence in E^* that is weak-* convergent is bounded.

(c) Show that weakly compact subsets of E and weak-* compact subsets of E^* are bounded.

1.20. Let X be a finite measure space. Via multiplication operators, we have an identification $L^\infty(X) \subset B(L^2(X))$. Show that the weak operator topology on $L^\infty(X)$ is the same as the weak-* topology (as the dual of $L^1(X)$.)

1.21. Let E be a Hilbert space and $U(E)$ the group of unitary operators. Show that the strong operator topology and the weak operator topology are the same on $U(E)$.

1.22. (a) Suppose $\{E_i\}$ are Hilbert spaces. Define $E = \{(f_1, f_2, \dots) \mid f_i \in E_i$ and $\sum \|f_i\|^2 < \infty\}$. Show that E is a Hilbert space (usually denoted by $\Sigma^\oplus E_i$). If $\{f_{ij}\}_j \subset E_i$ is an orthonormal basis, show $\{f_{ij}\}_{i,j}$ is an orthonormal basis of E, where we identify $E_i \subset E$ in the obvious way.

(b) Suppose E is a Hilbert space, and $E_i \subset E$ is a closed subspace. Suppose $E_i \perp E_j$ for $i \neq j$, and that the linear span of $\cup_j E_i$ is dense in E. Show $E \cong \Sigma^\oplus E_i$.

(c) If (X, μ) is a measure space which is expressed as a countable disjoint union $(X, \mu) = \cup(X_i, \mu_i)$, show $L^2(X)$ is naturally isomorphic to $\Sigma^\oplus L^2(X_i)$.

1.23 Prove the assertion in Example 1.1.19.

1.24 Prove the assertion in Example 1.3.14(b).

1.25 Let $\Omega \subset \mathbf{R}^n$ be open and $u \in C^0(\Omega)$. Suppose for all $|\alpha| \leq k$ that $D_w^\alpha u$ exists and $D_w^\alpha u \in C^0(\Omega)$. Show $u \in C^k(\Omega)$.

1.26 Suppose D_i, $i = 1, 2$, is a differential operator of order $\leq r_i$.

a) Show $D_1 \circ D_2$ is a differential operator of order $\leq r_1 + r_2$.

b) Show $D_1 \circ D_2 - D_2 \circ D_1$ is a differential operator of order $\leq (r_1 + r_2 - 1)$.

CONVEXITY AND FIXED POINT THEOREMS

2.1. Kakutani-Markov fixed point theorem

In the next two sections we describe fixed point theorems which in particular will yield results about the existence of invariant finite measures for certain group actions.

DEFINITION 2.1.1. *If a group G acts on a space X a point $x \in X$ is called a fixed point if $gx = x$ for all g.*

We recall that an action of the group $\mathbf{Z}$ on X is specified by giving a single invertible map $\varphi \colon X \to X$. In this case, there are some fixed point theorems known by topological methods. For example, the Brouwer fixed point theorem asserts that any continuous map $\varphi \colon X \to X$ has a fixed point if X is homeomorphic to the closed ball in $\mathbf{R}^n$. It is clear, however, that one may have homeomorphisms of other spaces with no fixed points. For example, a non-trivial rotation of the circle clearly has no fixed points. Any fixed point of course defines an invariant measure by taking the point mass at the fixed point. On the other hand, rotation on the circle, while having no fixed points, clearly leaves the arc length measure invariant. The Kakutani-Markov theorem, which is the main goal of this section, implies that every homeomorphism of a compact space has an invariant measure. There are two main ingredients. First is the compactness of $M(X)$ in the weak-*-topology (Corollary 1.1.29). The other is convexity.

DEFINITION 2.1.2. *If E is a vector space, a set $A \subset E$ is called convex if $x, y \in A$, $t \in [0, 1]$ implies $tx + (1 - t)y \in A$.*

EXAMPLE 2.1.3: (a) If E is a vector space and $\| \, \|$ is a seminorm on E, then open or closed $\| \, \|$-balls around any point x_0 (i.e. $\{x \mid \|x - x_0\| < r\}$ or $\{x \mid \|x - x_0\| \le r\}$) are convex.
(b) If X is a compact space, $M(X) \subset C(X)^*$ is convex.

(c) If $\lambda \in E^*$ (and $k = \mathbb{R}$), then for any $r \in \mathbb{R}$, $\{x \in E \mid \lambda(x) < r\}$ and $\{x \mid \lambda(x) \leq r\}$ are convex.

(d) If $\{A_\alpha\}$ are convex sets, so is $\bigcap A_\alpha$.

(e) If E is a TVS and $A \subset E$ is convex, so is $\overline{A}$.

DEFINITION 2.1.4. (a) If $A \subset E$, we define the convex hull of A, denoted by $\text{co}(A)$, to be the unique smallest convex set containing A. This exists by 2.1.3 (d) and is equal to $\bigcap \{B \subset E \mid A \subset B$ and B is convex$\}$.

(b) If E is a TVS, and $A \subset E$, we define the closed convex hull, denoted by $\overline{\text{co}}(A)$, to be the unique smallest closed convex set containing A. By 2.1.3 (d), (e), $\overline{\text{co}}(A) = \overline{\text{co}(A)} = \cap \{B \subset E \mid A \subset B$ and B is closed and convex$\}$.

THEOREM 2.1.5. (Kakutani-Markov) Let E be a TVS whose topology is defined by a sufficient family of seminorms. Suppose G is an abelian group and $\pi \colon G \to \text{Aut}(E)$ is a representation. Let $A \subset E$ be a compact convex set that is G-invariant, i.e. $\pi(g)A \subset A$ for all $g \in G$. Then there is a G-fixed point in A.

PROOF: For each $g \in G$ and $n \geq 0$, define $M_{n,g} \in B(E)$ by $M_{n,g} = \frac{1}{n} \sum_{i=0}^{n-1} \pi(g^i)$. Since A is convex and G-invariant, we have $M_{n,g}(A) \subset A$ for all n, g. Let G^* be the semigroup of operators generated $\{M_{n,g} \mid n \geq 0, g \in G\}$, (i.e. all finite compositions of such operators). Since G is abelian, G^* is commutative and we clearly have $T(A) \subset A$ for all $T \in G^*$. We claim $\bigcap_{T \in G^*} T(A) \neq \phi$, and that every element is a G-fixed point. To see the intersection is non-empty, since each $T(A)$ is compact (since A is compact and T is continuous), it suffices to see that for any finite set $T_1, \ldots, T_n \in G^*$, $\bigcap_{i=1}^n T_i(A) \neq \phi$. However, if we let $S = T_1 \circ \cdots \circ T_n \in G^*$, then $S(A) \subset T_1(T_2 \circ \cdots \circ T_n(A)) \subset T_1(A)$. Since G^* is commutative, we also have $S = T_2 \circ T_1 \circ \cdots \circ T_n$, and hence $S(A) \subset T_2(A)$. Similarly, $S(A) \subset T_i(A)$ for each i, showing that $\phi \neq S(A) \subset \bigcap T_i(A)$. Now suppose $y \in \bigcap_{T \in G^*} T(A)$. Then for each $n \geq 0$ and $g \in G$, there is some $x \in A$ such that $y = \frac{1}{n}(x + \cdots + \pi(g^{n-1})x)$. Then $\pi(g)y - y = (\pi(g^n)x - x)/n$. Let $\| \ \|$ be one of the seminorms defining the topology. Then for each n we have $\|\pi(g)y - y\| \leq 2B/n$ where $B = \sup\{\|a\| \mid a \in A\}$. (This exists since A is compact and $\| \ \| \colon E \to \mathbb{R}$ is continuous by definition.) Since this is true for all

n, $\|\pi(g)y - y\| = 0$, and since this is true for all seminorms in a sufficient family, $\pi(g)y = y$ for any $g \in G$.

COROLLARY 2.1.6. *Let G be an abelian group acting continuously on a compact metric space X. Then there is a G- invariant probability measure on X.*

PROOF: $M(X) \subset C(X)^*$ is compact, convex with the weak-*-topology. By 1.3.13, 1.3.14 we have a representation of G on $C(X)^*$ leaving $M(X)$ invariant. Thus, Theorem 2.1.5 implies the result.

EXAMPLE 2.1.7: It is not true that any group acting on a compact metric space has an invariant measure. For example, if we let $\varphi_1 \colon [0,1] \to [0,1]$ be $\varphi_1(x) = x^2$, the only invariant probability measures are supported on $\{0,1\}$. (See exercise 2.9.) Thus, if we identify φ_1 with a homeomorphism of S^1 by identifying 0 and 1, we obtain a homeomorphism whose only invariant measure is supported at a given point x_0. Let φ_2 be any homeomorphism of S^1 moving x_0, e.g. a rotation. Then the (non-abelian) group generated by φ_1, φ_2 has no invariant measure on S^1.

2.2. Haar measure for compact groups

Let G be a topological group. Then G acts on itself by left (or right) translation. I.e. for $g \in G$, we define the action of g on G to be $g \cdot h = gh$, where gh is simply multiplication. Under the assumption that G is locally compact, the following theorem asserts that there is always an essentially unique invariant measure. This result is fundamental for many aspects of the study of such groups.

THEOREM 2.2.1. (Haar) *Let G be a locally compact (second countable) group. Then:*

(i) *There is a measure μ which is invariant under left translations and is finite on compact subsets.*

(ii) *μ is unique up to positive scalar multiple.*

(iii) *The measure class of μ is the unique invariant measure class. More precisely if ν is a measure such that $g_*\nu \sim \nu$ (i.e. they have the same null sets), then $\nu \sim \mu$.*

(iv) *$\mu(G) < \infty$ if and only if G is compact.*

The measure μ is called the (left) Haar measure on G. If G is compact, we usually normalize so that $\mu(G) = 1$.

We shall not prove this result in general, although we shall prove it for G compact. In the compact case, this was first established by von Neumann. Here we shall see it as a consequence of a simple convexity argument (due to Kakutani.) We first discuss some examples.

EXAMPLE 2.2.2: (a) For $G = \mathbb{R}^n$, Haar measure is Lebesgue measure.

(b) For $G = S^1$, Haar measure is arc length. For the torus $T^n = (S^1)^n$, Haar measure is the product of the arc length measures on the factors.

(c) If G is discrete, Haar measure is counting measure.

(d) (This example requires some knowledge of manifolds and is not used in the sequel.) Suppose M is an oriented n- dimensional manifold. We let ω be a non-vanishing section of $\det(TM) = \Lambda^n(T^*M) \to M$, the top exterior power of the cotangent bundle of M. Then ω determines in a canonical way a measure μ_ω on M. If a group G acts on M, it acts on sections of $\det(TM)$, and ω will be G-invariant if and only if μ_ω is G-invariant. Now let $M = G$ be a Lie group. To construct an invariant measure on G, we need only construct an invariant section of $\det(TM)$. However, a basic argument of Lie theory shows that such invariant sections exist and in fact are in bijective correspondence with $\Lambda^n(T^*M)_e$ where $e \in G$ is the identity. Thus, the existence of an invariant measure for Lie groups follows easily from basic constructions in the theory of manifolds.

We now turn to the proof of the existence of an invariant probability measure for compact groups. We shall argue in a framework similar to that of Section 2.1. Namely, if X is a compact space, we shall realize $M(X) \subset C(X)^*$ as a weak-* compact convex set, and produce a fixed point in $M(X)$ for a compact group action.

THEOREM 2.2.3. *Let E be a Banach space and G be a compact group. Let $\pi: G \to \mathrm{Iso}(E)$ be a continuous isometric representation of G (where $\mathrm{Iso}(E)$ has the strong operator topology). Let $A \subset E_1^*$ be a compact convex G-invariant subset (for the adjoint*

π^* of π acting on E^*) where E_1^* has the weak-*-topology. Then there is a G-fixed point in A.

Compactness of G will be used only in the following lemma we need for the proof. It says, roughly, that we can take one of the seminorms on E^* defining the weak-*-topology and make it G-invariant.

LEMMA 2.2.4. Fix any $x \in E$. Define $\|\lambda\|_0$ for $\lambda \in E^*$ by $\|\lambda\|_0 = \sup\{|\lambda(\pi(g)x)| \mid g \in G\}$. Then

(i) $\| \ \|_0$ is a semi-norm on E^*.

(ii) $\| \ \|_0$ is G-invariant, i.e. for all $\lambda \in E^*, h \in G$, we have $\|\pi^*(h)\lambda\|_0 = \|\lambda\|_0$.

(iii) For any $0 \le r < \infty, \| \ \|_0 : E^* \to \mathbf{R}$ is continuous on E_r^* with the weak-*- topology.

(iv) If $\lambda(x) \ne 0$, then $\|\lambda\|_0 > 0$.

PROOF: We first remark that since G is compact and π is continuous, the map $g \mapsto \pi(g)x$ has compact image in E. Since λ is continuous on E, $\|\lambda\|_0$ is a well- defined real number. Then (i), (ii), and (iv) are straightforward from the definitions. To see (iii), suppose $\lambda_\alpha \to \lambda$ in the weak-*-topology where $\|\lambda_\alpha\|, \|\lambda\| \le r$. We claim $\|\lambda_\alpha - \lambda\|_0 \to 0$. Letting $\beta_\alpha = \lambda_\alpha - \lambda$, it suffices to see $\|\beta_\alpha\|_0 \to 0$, assuming $\beta_\alpha \to 0$ in the weak-*- topology and $\|\beta_\alpha\| \le 2r$ for all α. Fix $\varepsilon > 0$. Since $\{\pi(g)x \mid g \in G\}$ is compact, we can choose a finite $\varepsilon/4r$-dense subset $\{\pi(g_i)x \mid i = 1, \dots, n\}$. Since $\beta_\alpha \to 0$ in weak-*, for α sufficiently large we have $|\beta_\alpha(\pi(g_i)x)| < \varepsilon/2$ for all $i, 1 \le i \le n$. Then for any $g \in G$ we have for all i

$$|\beta_\alpha(\pi(g)x)| \le |\beta_\alpha(\pi(g)x - \pi(g_i)(x))| + |\beta_\alpha(\pi(g_i)x)|$$
$$\le 2r\|\pi(g)x - \pi(g_i)x\| + \varepsilon/2.$$

Choosing i so that $\|\pi(g)x - \pi(g_i)x\| < \varepsilon/4r$, we have $|\beta_\alpha(\pi(g)x)| \le \varepsilon$ for all $g \in G$, and hence $\|\beta_\alpha\|_0 \le \varepsilon$ as required.

PROOF OF THEOREM 2.2.3: We can apply Zorn's lemma to $\{B \subset A \mid B$ is compact, convex, G-invariant$\}$ with the ordering $B \ge C$ if and only if $B \subset C$. We deduce that there is a minimal G-invariant compact convex subset of A. Replacing A by this set, we may thus assume that A contains no proper convex compact G-invariant subset. We then wish to deduce that A consists of one point.

Suppose not. Then we can find $x \in E$ such that $\lambda_1(x) \neq \lambda_2(x)$ for some $\lambda_1, \lambda_2 \in A$. Using this x, form the semi-norm $\| \ \|_0$ as in Lemma 2.2.4. For $\lambda \in A$ and $r > 0$, we let $B(\lambda; r) = \{\beta \in A \mid \|\lambda - \beta\|_0 < r\}$. By Lemma 2.2.4, these are all open sets in A. We also set, for $r \geq 0$, $\overline{B}(\lambda; r) = \{\beta \in A \mid \|\lambda - \beta\|_0 \leq r\}$. These are all closed convex subsets. Let d be the $\| \ \|_0$-diameter of A, i.e. $d = \sup\{\|\lambda - \beta\|_0 \mid \lambda, \beta \in A\}$. Since A is compact and $\| \ \|_0$ is continuous on A, $d < \infty$. By our choice of x (used to define $\| \ \|_0$), we have $0 < d < \infty$. We claim that it suffices to find $\omega \in A$ and $r < d$ such that $\overline{B}(\omega; r) = A$. To see this, we simply observe that for such an ω, we have $\omega \in \overline{B}(\lambda; r)$ for all $\lambda \in A$, and hence $B = \bigcap_{\lambda \in A} \overline{B}(\lambda; r) \neq \phi$. But $B \subset A$ is clearly compact and convex, and we have $B \neq A$ since $r < d$. To obtain a contradiction to the minimality property of A, it only suffices to see that B is G-invariant. However, since $\| \ \|_0$ is G-invariant, $\overline{B}(\pi^*(g)\lambda; r) = \pi^*(g)(\overline{B}(\lambda; r))$, and hence $\pi^*(g)B = B$ for all $g \in G$.

To complete the proof, it therefore suffices to construct such an $\omega \in A$ and $r < d$. Since A is compact, we can find a finite set $\omega_1, \ldots, \omega_n \in A$ such that $A = \bigcup_{i=1}^n B(\omega_i; d/2)$. Let $\omega = \frac{1}{n} \sum_{i=1}^n \omega_i$. We have $\omega \in A$ since A is convex. Then for any $\lambda \in A$ we have

$$\|\omega - \lambda\|_0 \leq \frac{1}{n} \sum_{i=1}^n \|\omega_i - \lambda\|_0.$$

Each $\|\omega_i - \lambda\|_0 \leq d$ by the definition of d, but for at least one i we have $\|\omega_i - \lambda\|_0 \leq d/2$. Thus,

$$\|\omega - \lambda\|_0 \leq \frac{1}{n}\left(\frac{d}{2} + (n-1)d\right) = \left(\frac{2n-1}{2n}\right)d.$$

Therefore, letting $r = \frac{(2n-1)}{2n}d$ completes the proof.

COROLLARY 2.2.5. (von Neumann) *If G is any compact group, and X is any compact space on which G acts continuously, then there is a G- invariant probability measure on X. In particular, there is a probability measure on G invariant under left translations.*

PROOF: This follows from Theorem 2.2.3, using Proposition 1.3.9.

We now turn to the uniqueness assertion (ii) of Theorem 2.2.1 for G compact. It will be convenient at the same time to establish bi-invariance. Namely, we have that μ is left-invariant if $\mu(gA) = \mu(A)$ for all g, A. We can define μ to be right invariant if $\mu(Ag) = A$ for all g, A. We call μ bi-invariant if it is both left and right invariant.

PROPOSITION 2.2.6. *Let G be a compact group, and $\mu \in M(G)$ with μ left invariant. Then μ is the unique left invariant probability measure, and μ is bi-invariant.*

PROOF: Let ν be a right invariant probability measure on G. (This exists either by appeal to Theorem 2.2.3 applied to G acting on itself by right translations, or by taking $\nu = I_*\mu$, where $I(g) = g^{-1}$.) Then for any $f \in C(G)$, we have (by right invariance of ν)

$$\int f(yx)d\mu(x)d\nu(y) = \int \left(\int f(y)d\nu(y) \right) d\mu(x) = \int fd\nu.$$

Similarly, by left invariance of μ,

$$\int f(yx)d\mu(x)d\nu(y) = \int fd\mu.$$

Thus $\mu = \nu$, showing bi-invariance. Since $\mu' = \nu$ for any other left invariant $\mu' \in M(G)$, we have $\mu = \mu'$, showing uniqueness.

REMARK 2.2.7: It is not true that left Haar measure on a general locally compact group is bi-invariant. A group for which this holds is called unimodular. For an example of a non-unimodular group see exercise 3.12.

We now indicate some first applications of the existence of Haar measure for compact groups.

COROLLARY 2.2.8. *Let G be a compact group and V a finite dimensional vector space. Let $\pi\colon G \to GL(V)$ be a continuous representation. Then there is a (positive definite) inner product $\langle\ ,\ \rangle$ on V which is G - invariant, i.e. $\langle\pi(g)v, \pi(g)w\rangle = \langle v, w\rangle$ for all $v, w \in V$. In other words π is orthogonal (resp. unitary) if $k = \mathbb{R}$ (resp. $\mathbb{C}$).*

PROOF: Let $(\ ,\ )$ be any inner product on V. Let

$$\langle v, w\rangle = \int_G (\pi(g)v, \pi(g)w)d\mu(g)$$

where μ is Haar measure. Then one easily checks that $\langle\ ,\ \rangle$ satisfies the required conditions.

As a consequence of Corollary 2.2.8 we obtain complete reducibility of a finite dimensional representation of a compact group.

DEFINITION 2.2.9. *Suppose V is a finite dimensional vector space and $\pi\colon G \to GL(V)$ is a representation. Then π is called irreducible if the only $\pi(G)$-invariant subspaces are (0) and V. π is called completely reducible if $V = \sum^{\oplus} V_i$, where V_i is $\pi(G)$-invariant and irreducible.*

EXAMPLE 2.2.10: For $t \in \mathbb{R}$, let $\pi(t) = \begin{pmatrix} 1 & t \\ 0 & 1 \end{pmatrix}$. Thus $\pi\colon \mathbb{R} \to GL(2, \mathbb{R})$ is a representation. The space $\mathbb{R}e_1 \subset \mathbb{R}^2$ is invariant and irreducible (since it is 1- dimensional.) But π is not completely reducible.

PROPOSITION 2.2.11. *Let G be a group and $\pi\colon G \to U(n)$ be a (finite dimensional) unitary representation. Then*
  (i) *every $\pi(G)$- invariant subspace has a $\pi(G)$- invariant complement.*
  (ii) *π is completely reducible.*

PROOF: If V is invariant and π is unitary then $V^{\perp}$ is invariant. Namely, if $w \in V^{\perp}$ and $g \in G$, then for all $v \in V$ we have $\langle\pi(g)w, v\rangle = \langle w, \pi(g)^*v\rangle = \langle w, \pi(g^{-1})v\rangle$. This is 0 since

$\pi(g^{-1})v \in V$. Since $V \oplus V^{\perp} = \mathbf{C}^n$, this proves (i). To see (ii), simply argue by induction on n. Namely, if π is not irreducible, write $\mathbf{C}^n = V_1 \oplus V_2$ where V_i are $\pi(G)$-invariant and dim $V_i <$ dim V. Write $\pi(g) = \pi_1(g) \oplus \pi_2(g)$ where $\pi_i(g) = \pi(g) \mid V_i$. Then apply the induction hypothesis to π_i.

The same argument clearly works for orthogonal representations over $\mathbf{R}$.

COROLLARY 2.2.12. *If G is a compact group, then any continuous representation $\pi: G \to GL(V)$, with dim $V < \infty$, is completely reducible.*

PROOF: Proposition 2.2.11 and Corollary 2.2.8.

We shall discuss the case of infinite dimensional representations in Section 3.3. We conclude this section with some examples of compact groups.

EXAMPLE 2.2.13: (a) We have already mentioned $O(n) \subset GL(n, \mathbf{R})$ and $U(n) \subset GL(n, \mathbf{C})$ as examples. In fact any compact subgroup of $GL(n, \mathbf{R})$ (resp. $GL(n, \mathbf{C})$) is conjugate to a closed subgroup of $O(n)$ (resp. $U(n)$). Namely, if $G \subset GL(n, \mathbf{R})$ is compact, then by Corollary 2.2.8 there is a positive definite inner product on $\mathbf{R}^n$ such that G leaves the inner product invariant. Let $v_1, \ldots, v_n$ be an orthonormal basis with respect to this inner product, and let $T \in GL(n, \mathbf{R})$ be given by $Te_i = v_i$. Then $T^{-1}GT$ leaves the standard inner product invariant, i.e. $T^{-1}GT \subset O(n)$. The argument over $\mathbf{C}$ is similar.

(b) Any finite group is compact. Hence any (possibly infinite) product of finite groups is compact. Furthermore, any closed subgroup of a product of finite groups is compact. (There are a number of groups that arise naturally in an algebraic setting that have natural realizations of this form, for example, the p-adic integers, or the Galois group of an infinite Galois extension.)

2.3. Krein-Millman Theorem

The Krein-Millman theorem concerns a geometric feature of a compact convex set A, which allows one to describe A in terms of a certain natural subset of A.

DEFINITION 2.3.1. *If E is a vector space and $x, y \in E$, we let $[x, y] = \{tx + (1 - t)y \mid t \in [0, 1]\}$ and $(x, y) = \{tx + (1 - t)y \mid t \in (0, 1)\}$. Let $A \subset E$ be a convex set. If $x \in A$, we say that x is an extreme point of A if $x \in [y, z]$ for some $y, z \in A$ implies $x = y$ or $x = z$. Equivalently, $x \in (y, z)$ implies $x = y = z$. More generally, if $\emptyset \neq B \subset A$ is convex, we say that B is an extreme set in A if $y, z \in A$ and $(y, z) \cap B \neq \emptyset$, implies $[y, z] \subset B$. (Thus $\{x\}$ is an extreme set if and only if x is an extreme point.) We let $ex(A)$ be the set of extreme points of A.*

EXAMPLE 2.3.2: (a) In $\mathbb{R}^n$, the extreme points of a closed convex polyhedron are just the vertices. Every edge, face, etc. is an extreme set, although there are other extreme sets.
(b) The set of extreme points of a closed ball in $\mathbb{R}^n$ (usual inner product) is precisely the boundary sphere.

EXAMPLE 2.3.3: Let X be a compact metric space. Then $ex(M(X)) = \{\delta_x \mid x \in X\}$. (See exercise 2.10.)

THEOREM 2.3.4. (Krein-Millman) *Let E be a TVS whose topology is defined by a sufficient family of seminorms. If $A \subset E$ is compact, convex, then $\overline{co}(ex(A)) = A$. (We recall that $\overline{co}(\ )$ denotes the closed convex hull [Definition 2.1.4].)*

For the proof, we need the following generalized version of Hahn-Banach.

LEMMA 2.3.5. *Suppose E is a real TVS whose topology is defined by a sufficient family of seminorms. Let $A \subset E$ be closed and convex, and $x \in E$ and $x \notin A$. Then there is some $\lambda \in E^*$ and $\alpha \in \mathbb{R}$ such that $\lambda(a) < \alpha < \lambda(x)$ for all $a \in A$.*

The proof of Lemma 2.3.5 is indicated in exercise 2.1.

PROOF OF THEOREM 2.3.4: By restricting the field of scalars if necessary, we may assume E is a real vector space. We claim first that every closed convex extreme subset B of A contains an extreme point of A. To see this, we observe that we can apply Zorn's lemma to the collection of closed convex subsets C of B which are extreme in A (ordered by reverse inclusion) to show that there is a minimal such subset. We claim that C consists of a single point, i.e. is an extreme point. If we have $x, y \in C, x \neq y$, by Hahn-Banach we can find $f \in E^*$ such $f(x) < f(y)$. Let $M = \max\{f(z) \mid z \in C\}$ which exists since C is compact. Let $D = \{w \in C \mid f(w) = M\}$. Then D is an extreme set of A, $D \subset C$, and $D \neq C$ since $x \notin D$. This contradicts the minimality of C, showing the existence of extreme points in B.

Now suppose $A \neq \overline{co}(\mathrm{ex}(A))$. Note that we clearly have $A \supset \overline{co}(\mathrm{ex}(A))$. Thus, we may choose $x \in A, x \notin \overline{co}(\mathrm{ex}(A))$. By Lemma 2.3.5 we can find $\alpha \in \mathbf{R}$ and $\lambda \in E^*$ such that $\lambda(y) < \alpha < \lambda(x)$ for all $y \in \overline{co}(\mathrm{ex}(A))$. As above, we let M be the maximum value of λ on A, and let $B = \{z \in A \mid \lambda(z) = M\}$. Then B is an extreme subset of A, and since $M > \alpha$, we have $B \cap \overline{co}(\mathrm{ex}(A)) \neq \phi$. However, by the preceding paragraph, B contains an extreme point of A, which is a contradiction.

COROLLARY 2.3.6. *If E is a TVS whose topology is defined by a sufficient family of semi-norms, and $A \subset E$ is compact, convex (and $\neq \phi$), then $\mathrm{ex}(A) \neq \phi$.*

As an application of 2.3.6, we consider invariant measures.

DEFINITION 2.3.7. *Suppose a topological group G acts continuously on a compact metric space X. Let $\mu \in M(X)$ be G-invariant. We say that μ is ergodic (for the G-action) if $A \subset X$ is a measurable G- invariant set (i.e. $gA = A$ for all $g \in G$) implies $\mu(A) = 0$ or $\mu(A) = 1$.*

EXAMPLE 2.3.8: (a) Let $X = S^1 = \{z \in \mathbf{C} \mid |z| = 1\}$, and $\mu = $ normalized arc length. Suppose G is S^1 acting on itself by multiplication. Thus, for $z = e^{i\alpha}$, z acts on S^1 by rotation by angle α. Clearly the only invariant sets (measurable or not) are S^1 and ϕ, hence μ is ergodic.
(b) With X as above, fix $\alpha \in \mathbf{R}$. Consider the $\mathbf{Z}$-action generated by the homeomorphism $\phi(z) = e^{i\alpha}z$. Then μ is not necessarily ergodic for this $\mathbf{Z}$-action. Namely, if $\alpha = \pi$, then $\{e^{i\theta} \mid \theta \in (0, \pi/2) \cup (\pi, 3\pi/2)\}$ is clearly invariant and of measure $1/2$. A simple modification of the argument shows that μ will not be ergodic for the $\mathbf{Z}$-action defined by any α with $\alpha/2\pi \in \mathbf{Q}$. On the other hand, if $\alpha/2\pi \notin \mathbf{Q}$, then μ will be ergodic. This can be seen for example by a Fourier series argument. (See exercise 2.13).

PROPOSITION 2.3.9. *If a topological group G acts continuously on a compact metric space X, let $M(X)^G \subset M(X)$ be the set of G-invariant measures. Then $M(X)^G$ is compact and convex. If $M(X)^G \neq \phi$, then any $\mu \in ex(M(X)^G)$ is ergodic.*

PROOF: The first assertion is immediate. If μ is not ergodic, let $A \subset X$ be invariant and measurable with $0 < \mu(A) < 1$. For $B \subset X$ with $\mu(B) > 0$, let $\mu_B \in M(X)$ be $\mu_B(Z) = \mu(Z \cap B)/\mu(B)$. Since A, and hence $X - A$, are G-invariant, we have $\mu_A, \mu_{X-A} \in M(X)^G$, and $\mu = t\mu_A + (1-t)\mu_{X-A}$ where $t = \mu(A)$. Thus, μ is not extreme in $M(X)^G$.

COROLLARY 2.3.10. *If G is an abelian group acting continuously on a compact metric X, then there is an ergodic, G-invariant $\mu \in M(X)$.*

PROOF: This follows from the Kakutani-Markov theorem, Proposition 2.3.9, and Corollary 2.3.6.

PROBLEMS FOR CHAPTER 2

2.1. Prove the following version of the Hahn-Banach theorem. Let E be a LCTVS, $A \subset E$ closed and convex, and $x \in E$ with $x \in A$. Show there is some $f \in E^*$ and $c \in \mathbf{R}$ such that $\mathrm{Re}(f(y)) < c < \mathrm{Re}(f(x))$ for all $y \in A$.

Hint: Choose a convex balanced U such that $(x + U) \cap A = \phi$. If $0 \in A$, consider the seminorm defined by $A + U/2$. (cf. Problem 1.6.)

2.2. Suppose $A \subset B$ are compact sets in an LCTVS, with B convex. Show that every extreme point of $\overline{co}(A)$ lies in A.

2.3. If (X, μ) is a measure space, a mean on $L^\infty(X)$ is a norm continuous linear functional $m: L^\infty(X) \to \mathbf{C}$ such that

(i) $m(1) = 1$.

(ii) $f \geq 0$ implies $m(f) \geq 0$.

(iii) $m(\overline{f}) = \overline{m(f)}$.

(a) Let $\mathcal{M}(X)$ be the set of means on $L^\infty(X)$. Show $\mathcal{M}(X)$ is a closed convex subset of the unit ball in $L^\infty(X)^*$.

(b) Let $X = \mathbf{R}^n$ with Lebesgue measure. Then $m \in \mathcal{M}(\mathbf{R}^n)$ is called invariant if for all $f \in L^\infty(\mathbf{R}^n)$ and all $t \in \mathbf{R}^n$ we have $m(f) = m(\pi(t)f)$ where $(\pi(t)f)(x) = f(x - t)$. Show that an invariant mean on $\mathbf{R}^n$ exists.

(c) If m is an invariant mean and $A \subset \mathbf{R}^n$ is bounded, show $m(\chi_A) = 0$.

(d) If m is an invariant mean and $f \in L^\infty(\mathbf{R}^n)$ vanishes off a compact set, show $m(f) = 0$.

(e) If m is an invariant mean, show $m: L^\infty(\mathbf{R}^n) \to \mathbf{C}$ is not weak-*-continuous.

2.4. Define invariant means on $\mathbf{Z}^n$ and derive results analogous to those of problem 2.3.

2.5. Let E be a Banach space and suppose $T \in \mathrm{Aut}(E)$ satisfies the condition $\|T^n\| \leq C$ for all $n \in \mathbf{Z}$ (where $C \in \mathbf{R}$). Show there is an equivalent norm on E, say $|\ |$, such that T is isometric with respect to $|\ |$. Hint: For $x \in E$, define $f(n) = \|T^n x\|$. Then $f \in \ell^\infty(\mathbf{Z})$. Apply problem 2.4.

2.6. If G is a compact group, μ is Haar measure, and $U \subset G$ is open, show $\mu(U) > 0$.

2.7. Let E be a TVS which is either a Banach space with the weak topology or the dual of a Banach space with the weak-*-

topology.

(a) Let $A \subset E$ be compact. Let $\mu \in M(A)$, i.e. μ is a probability measure on A. Show there is a unique point $b(\mu) \in E$ such that for all $\lambda \in E^*, \lambda(b(\mu)) = \int_A \lambda(a) d\mu(a)$. ($b(\mu)$ is called the barycenter of μ.) What is $b(\mu)$ if μ is supported on a finite set? Hint: Use Proposition 1.1.30.

(b) Show the map $\mu \mapsto b(\mu)$ defines a continuous map $M(A) \to E$ where $M(A)$ has the weak-*-topology. (You may want to use problem 1.19.)

(c) Show that $b(M(A))$ is precisely $\overline{co}(A)$.

(d) Show that $\overline{co}(A)$ is compact.

2.8. (a) Let $A \subset \mathbb{R}^2$ be a circle. Show that the map $b \colon M(A) \to \overline{co}(A)$ is not injective.

(b) Let $A \subset \mathbb{R}^2$ consist of three non-colinear points. Show that $b \colon M(A) \to \overline{co}(A)$ is a homeomorphism.

2.9. Let $I = [0,1]$ and $f \colon I \to I$ be $f(x) = x^2$. Find all f- invariant probability measures on I.

2.10 If X is a compact Hausdorff space, show $ex(M(X)) = \{\delta_x \mid x \in X\}$ where δ_x is the probability measure supported on $\{x\}$.

2.11 If E is a Banach space, G a compact group and $\pi \colon G \to \mathrm{Aut}(E)$ is a continuous representation (for the strong operator topology), show there is an equivalent norm on E for which π is isometric. (Cf. problem 2.5). If E is a Hilbert space, show that this norm can be taken to come from an inner product on E.

2.12 Suppose G is a compact group and G acts continuously on a compact metrizable space X. Show there is a metric on X (defining the same topology) that is G-invariant, i.e. $d(gx, gy) = d(x, y)$ for all $x, y \in X$, and all $g \in G$. (Hint: Let ρ be any metric. Then average.)

2.13 (a) Suppose $\alpha \in \mathbb{R}$, $\alpha/2\pi \notin \mathbb{Q}$. Suppose $f \in L^2(S^1)$ with $f(e^{i\alpha}z) = f(z)$ for all z. Let $f(z) = \sum_{-\infty}^{\infty} a_n z^n$ be the L^2-Fourier expansion of f. Show $a_n = 0$ if $n \neq 0$.

(b) Verify the ergodicity assertion in Example 2.3.8(b).

COMPACT OPERATORS

3.1. Compact operators and Hilbert-Schmidt operators

Certain natural classes of operators between Banach spaces have properties considerably stronger than boundedness.

DEFINITION 3.1.1. *If E, F are Banach spaces, a bounded operator $T: E \to F$ is called a compact operator if $\overline{T(E_1)}$ is compact in F. Equivalently, $\overline{T(B)}$ is compact for every bounded set B.*

EXAMPLE 3.1.2: If T has finite rank, i.e. $\dim T(E) < \infty$, then T is compact.

LEMMA 3.1.3. *Suppose $T_n \in B(E,F)$ are compact and $\|T_n - T\| \to 0$. Then T is compact.*

PROOF: It suffices to see that $T(E_1)$ is totally bounded, i.e. for any $\varepsilon > 0$, there is a finite ε-dense set. For any $x, y \in E_1$ and any n, we have

$$\|Tx - Ty\| \leq \|Tx - T_n x\| + \|T_n x - T_n y\| + \|T_n y - Ty\|$$
$$\leq 2\|T - T_n\| + \|T_n x - T_n y\|.$$

Fix n sufficiently large so that $\|T - T_n\| \leq \varepsilon/4$. Then choose a finite $\varepsilon/2$-dense set for $T_n(E_1)$, say $T_n(x_1), \ldots, T_n(x_r)$. It then follows that $T(x_1), \ldots, T(x_r)$ is ε-dense in $T(E_1)$.

EXAMPLE 3.1.4: Suppose E is a Hilbert space with orthonormal basis $\{e_i\}$ and $T \in B(E)$ is given by the diagonal matrix $T_{ij} = \lambda_i \delta_{ij}$. I.e. $Te_i = \lambda_i e_i$. Then T is compact if and only if $\lambda_i \to 0$ as $i \to \infty$; equivalently, for any $\varepsilon > 0, \{i \mid |\lambda_i| > \varepsilon\}$ is finite. To see this, first suppose $\lambda_i \to 0$. Then for each n, let $T_n e_i =$

Te_i if $i \leq n$ and $T_n e_i = 0$ if $i > n$. Then T_n has finite rank and hence is compact, and it is straightforward to see that $\|T - T_n\| = \sup\{|\lambda_i| \mid i > n\}$. (In fact this follows from 1.2.1 and 1.2.17.) Thus by 3.1.3, T is compact. Conversely, if there is some ε with $S = \{i \mid |\lambda_i| > \varepsilon\}$ infinite, then for $i, j \in S$ we have $\|Te_i - Te_j\|^2 = |\lambda_i|^2 + |\lambda_j|^2 \geq 2\varepsilon^2$. Therefore $\{Te_i\}_{i \in S}$ obviously has no convergent subsequence, showing that T is not compact.

The next result gives an important class of compact operators.

THEOREM 3.1.5. *Let X be a compact space with a finite measure μ. If $K \in C(X \times X)$, then the integral operator*

$$(T_K f)(X) = \int_X K(x, y) f(y) d\mu(y)$$

defines a compact operator $T_K : L^2(X) \to L^2(X)$.

REMARK: (A simple modification of the proof will show that the same formula defines a compact operator on $L^p(X)$ for any $1 \leq p < \infty$.)

PROOF: First suppose K is of the form

$$K(x, y) = \sum_{i=1}^{n} \varphi_i(x) \psi_i(y) \qquad \text{for } \varphi_i, \psi_i \in C(X).$$

Then

$$(T_K f)(x) = \sum_{i=1}^{n} \left(\int_X \psi_i f \right) \varphi_i(x),$$

so that

$$T_K(L^2(X)) \subset \text{span}\{\varphi_1, \ldots, \varphi_n\}.$$

In particular, T_K has finite rank and hence is compact. For a general K, we can find K_j of the above form with $K_j \to K$ uniformly. (Example A.9 (c).) Furthermore $T_{K_j} - T_K = T_{K_j - K}$ and hence $\|T_{K_j} - T_K\| \leq \|K_j - K\|_{L^2(X \times X)}$ (by Example 1.2.14.) Since $K_j \to K$ uniformly, $K_j \to K$ in $L^2(X \times X)$, and hence $\|T_{K_j} - T_K\| \to 0$. The result now follows from Lemma 3.1.3.

We shall shortly present another proof of 3.1.5, assuming only that $K \in L^2(X \times X)$.

DEFINITION 3.1.6. *Let E be a Hilbert space with orthonormal basis $\{e_i\}$. If $T \in B(E)$, then T is called a Hilbert-Schmidt operator if*

$$\sum_{i,j} |T_{ij}|^2 < \infty, \quad i.e. \quad \sum_{i,j} |\langle Te_j, e_i \rangle|^2 < \infty.$$

Equivalently $\sum_j \|Te_j\|^2 < \infty$.

While this appears to depend on the basis, we show it does not.

LEMMA 3.1.7. *If $\{e_i\}$ and $\{f_j\}$ are orthonormal bases, then*

$$\sum_{i,j} |\langle Te_j, e_i \rangle|^2 = \sum_{i,j} |\langle Tf_j, f_i \rangle|^2.$$

PROOF: We have

$$\sum_j \left(\sum_i |\langle Te_j, e_i \rangle|^2 \right) = \sum_j \|Te_j\|^2$$
$$= \sum_{j,i} |\langle Te_j, f_i \rangle|^2$$
$$= \sum_{j,i} |\langle e_j, T^* f_i \rangle|^2.$$

Thus,

$$\sum_{i,j} |\langle Te_j, e_i \rangle|^2 = \sum_j \|T^* f_i\|^2.$$

The right hand side is independent of the orthonormal basis $\{e_i\}$, and hence so is the left side.

DEFINITION 3.1.8. *If T is Hilbert-Schmidt, we define the Hilbert-Schmidt norm of T, $\|T\|_2$ by*

$$\|T\|_2 = \left(\sum_{i,j} |T_{ij}|^2 \right)^{1/2}$$

where T_{ij} is the matrix of T with respect to any orthonormal basis.

We remark that the proof of Lemma 3.1.7 shows:

COROLLARY 3.1.9. *If T is Hilbert-Schmidt, so is T^*, and $\|T\|_2 = \|T^*\|_2$.*

LEMMA 3.1.10. *If T is Hilbert-Schmidt then $\|T\| \leq \|T\|_2$.*

PROOF: Let $\{e_i\}$ be an orthonormal basis of E, and $x \in E$. Then $x = \sum c_j e_j$. Therefore

$$\|Tx\|^2 = \sum_i \left| \langle T(\sum_j c_j e_j), e_i \rangle \right|^2$$
$$= \sum_i \left| \sum_j c_j \langle T e_j, e_i \rangle \right|^2.$$

By Cauchy-Schwarz, this is

$$\leq \sum_i \left(\sum_j |c_j|^2 \right) \left(\sum_j |\langle T e_j, e_i \rangle|^2 \right)$$
$$\leq \|x\|^2 \sum_{i,j} |\langle T e_j, e_i \rangle|^2$$
$$\leq \|x\|^2 \|T\|_2^2.$$

This proves the lemma.

PROPOSITION 3.1.11. *If T is Hilbert-Schmidt, then T is compact.*

PROOF: Let $\{e_i\}$ be an orthonormal basis. For each n, define $T_n \in B(E)$ by $T_n e_i = e_i$ if $i \leq n$, $T_n e_i = 0$ if $i > n$. Then T_n has finite rank. Furthermore, $T - T_n$ is Hilbert-Schmidt with $\|T - T_n\|_2^2 = \sum_{i>n} \|T e_i\|^2$. Since $\sum_i \|T e_i\|^2 < \infty$, we have $\|T - T_n\|_2 \to 0$ as $n \to \infty$. By Lemma 3.1.10, $\|T - T_n\| \to 0$, and by 3.1.3, T is compact.

PROPOSITION 3.1.12. *Let X be a measure space and $K \in L^2(X \times X)$. Then $T_K : L^2(X) \to L^2(X)$ is Hilbert-Schmidt and hence compact. Furthermore, $\|T_K\|_2 = \|K\|_2$.*

PROOF: We let $K_x \in L^2(X)$ be $K_x(y) = K(x, y)$. (This is in $L^2(X)$ for a.e. x by Fubini's theorem.) Let $\{e_i\}$ be an orthonormal basis of $L^2(X)$. Then

$$\begin{aligned}
\sum_i \|T_K e_i\|^2 &= \sum_i \int_X |(T_K e_i)(x)|^2 \, dx \\
&= \sum_i \int_X |\langle K_x, \overline{e}_i \rangle|^2 \, dx \\
&= \int_X \left(\sum_i |\langle \overline{K}_x, e_i \rangle|^2 \right) dx \\
&= \int_X \|\overline{K}_x\|_2^2 \\
&= \|K\|_{L^2(X \times X)}^2
\end{aligned}$$

REMARK 3.1.13: a) Let $B_2(E)$ denote the space of Hilbert-Schmidt operators on E. With the Hilbert-Schmidt norm, $B_2(E)$ is actually a Hilbert space. Namely, if B is an orthonormal basis, then $B_2(E)$ can be identified with $\ell^2(B \times B)$. We have the inner product on $B_2(E)$ given by $\langle T, S \rangle = \sum_{i,j} T_{ij} \overline{S}_{ij}$, or equivalently, $\langle T, S \rangle = \sum_i (TS^*)_{ii}$.

b) One can form spaces analogous to $B_2(E)$ for any p, $1 \le p < \infty$, using an ℓ^p-type norm. The most important such operators are "trace class operators", where $p = 1$.

c) The Banach space $B(E)$ is also an algebra under composition, for any Banach space E. We have that the space of compact operators, which we denote by $B_c(E) \subset B(E)$, is a closed (by 3.1.3) linear subspace. In fact, $B_c(E)$ is a closed two sided ideal in the algebra $B(E)$. This follows from the following easy fact.

PROPOSITION 3.1.14. If $T : E \to F$, $S : F \to L$ are bounded linear maps between Banach spaces, then $S \circ T$ is compact if either S or T is compact.

3.2. Spectral theorem for compact operators

If E is a finite dimensional Hilbert space and $T \in B(E)$ is self-adjoint then the spectral theorem in finite dimensions asserts that T has an orthonormal basis of eigenvectors. In this section, we generalize this result to compact self-adjoint operators on a general Hilbert space.

DEFINITION 3.2.1. *Let E be a Banach space and $T \in B(E)$. For $\lambda \in \mathbb{C}$, let $E_\lambda = \{x \in E \mid Tx = \lambda x\}$. Then $E_\lambda \subset E$ is a closed linear subspace. λ is called an eigenvalue if $E_\lambda \neq (0)$, and $x \in E_\lambda$ is called an eigenvector with eigenvalue λ.*

EXAMPLE 3.2.2: In contrast to finite dimensions, a self-adjoint bounded operator need not have any eigenvalues. For example, let $T = M_x$, multiplication by the (L^∞) function $x \mapsto x$ on $L^2([0,1])$. Then $(M_x)^* = M_{\bar{x}} = M_x$, so that T is self-adjoint. However, if $Tf = \lambda f$ for some $f \in L^2([0,1])$, then $\lambda f(x) = xf(x)$ a.e., and hence $f = 0$ in $L^2([0,1])$.

The main result of this section is:

THEOREM 3.2.3. *(Spectral theorem for compact operators) Suppose E is a Hilbert space and $T \in B(E)$ is compact and self-adjoint. Then E has an orthonormal basis consisting of eigenvectors for T. Furthermore, for each $\lambda \neq 0$, $\dim E_\lambda < \infty$, and for each $\varepsilon > 0$, $\{\lambda \mid |\lambda| \geq \varepsilon \text{ and } \dim E_\lambda > 0\}$ is finite.*

Before proving the theorem, we collect some general facts about self-adjoint operators.

LEMMA 3.2.4. *Let E be a Hilbert space and $T \in B(E)$.*

  i) *If $T = T^*$ and W is T-invariant (i.e., $T(W) \subset W$), then $W^\perp$ is also T-invariant.*

  ii) *If $T = T^*$, then for all $x \in E$ we have $\langle Tx, x \rangle \in \mathbb{R}$. In particular, all eigenvalues are real.*

  iii) *$\|T\| = \sup\{|\langle Tx, y \rangle| \mid \|x\|, \|y\| \leq 1\}$.*

  iv) *If $T = T^*$, then $\|T\| = \sup\{|\langle Tx, x \rangle| \mid \|x\| \leq 1\}$.*

  v) *If $T = T^*$ and $\lambda \neq \beta$, then $E_\lambda \perp E_\beta$.*

PROOF: i) If $y \in W^\perp$, then for any $x \in W$, $\langle x, Ty \rangle = \langle Tx, y \rangle = 0$ since $Tx \in W$.

ii) $\overline{\langle Tx, x \rangle} = \langle x, Tx \rangle = \langle T^*x, x \rangle = \langle Tx, x \rangle$.

iii) Clearly $|\langle Tx, y \rangle| \leq \|T\|$ if $\|x\|, \|y\| \leq 1$. Conversely, assume $T \neq 0$. For any y with $Ty \neq 0$, we have for $x = Ty/\|Ty\|$ that $|\langle Ty, x \rangle| = \|Ty\|$. Taking the supremum over all y with $\|y\| = 1$ yields $\|T\|$.

iv) Let $\alpha = \sup\{|\langle Tx, x \rangle| \mid \|x\| \leq 1\}$. We clearly have $\alpha \leq \|T\|$.

By (iii), to prove (iv) it suffices to see $|\langle Tx, y \rangle| \leq \alpha \|x\| \, \|y\|$ for all x, y. As this inequality is unchanged if we multiply y by a complex number of modulus 1, we may assume $\langle Tx, y \rangle \in \mathbf{R}$. We now express $\langle Tx, y \rangle$ in terms of expressions of the form $\langle Tw, w \rangle$. Namely, we have

$$\langle T(x+y), (x+y) \rangle = \langle Tx, x \rangle + \langle Tx, y \rangle + \langle Ty, x \rangle + \langle Ty, y \rangle.$$

Since $T = T^*$ and $\langle Tx, y \rangle \in \mathbf{R}$, this reduces to

$$\langle T(x+y), (x+y) \rangle = \langle Tx, x \rangle + 2\langle Tx, y \rangle + \langle Ty, y \rangle.$$

Similarly,

$$\langle T(x-y), (x-y) \rangle = \langle Tx, x \rangle - 2\langle Tx, y \rangle + \langle Ty, y \rangle.$$

Subtracting the last equation from the one preceding it, we obtain

$$4\langle Tx, y \rangle = \langle T(x+y), x+y \rangle - \langle T(x-y), x-y \rangle$$

Therefore,

$$|\langle Tx, y \rangle| \leq \frac{\alpha}{4} \big(\|x+y\|^2 + \|x-y\|^2 \big).$$

By the parallelogram inequality, we have

$$\|x+y\|^2 + \|x-y\|^2 \leq 2 \big(\|x\|^2 + \|y\|^2 \big)$$

and thus

$$|\langle Tx, y \rangle| \leq \frac{\alpha}{2} \big(\|x\|^2 + \|y\|^2 \big)$$

For any $a > 0$, we apply this last equation to $\sqrt{a}x$, $\frac{1}{\sqrt{a}}y$. We thus have for all $x, y \in E$ and $a > 0$ that

$$|\langle Tx, y \rangle| \leq \frac{\alpha}{2} \big(a\|x\|^2 + \frac{1}{a}\|y\|^2 \big).$$

We clearly need only consider $\|x\| \neq 0$, and we now set $a = \|y\|/\|x\|$. We then obtain $|\langle Tx, y \rangle| \leq \alpha$ as required.

v) If $Tx = \lambda x$ and $Ty = \beta y$, then we have

$$\langle Tx, y \rangle = \lambda \langle x, y \rangle, \quad and \quad \langle Tx, y \rangle = \langle x, Ty \rangle = \beta \langle x, y \rangle.$$

Since $\lambda \neq \beta$, we have $\langle x, y \rangle = 0$.

PROOF OF THEOREM 3.2.3: We first claim that we can find at least one non-zero eigenvector (assuming, of course, that $E \neq (0)$). If $T = 0$, then any vector is an eigenvector, so we may assume $T \neq 0$. By Lemma 3.2.4(iv) we can choose $x_n \in E$ with $\|x_n\| = 1$ and $\langle Tx_n, x_n \rangle \to \lambda$ where $|\lambda| = \|T\|$. Since T is compact, by passing to a subsequence we may assume $Tx_n \to y$ for some $y \in E$. Since $\lambda \neq 0$, we must have $y \neq 0$ as well. We now observe that (using $T = T^*$ and 3.2.4(ii)),

$$\|Tx_n - \lambda x_n\|^2 = \|Tx_n\|^2 - 2\lambda \langle Tx_n, x_n \rangle + \lambda^2 \|x_n\|^2$$
$$\leq 2\|T\|^2 - 2\lambda \langle Tx_n, x_n \rangle.$$

As $n \to \infty$, we thus have $\|Tx_n - \lambda x_n\| \to 0$. Since $Tx_n \to y$, we thus have $\lambda x_n \to y$ as well. Applying T to this last convergence we have $\lambda Tx_n \to Ty$. But we also have from $Tx_n \to y$ that $\lambda Tx_n \to \lambda y$. Hence $Ty = \lambda y$, and hence y is an eigenvector.

We now complete the proof. By Zorn's lemma we can choose an orthonormal set of eigenvectors for T which is maximal among all orthonormal sets of eigenvectors. Let W be the closure of the span of these vectors. It suffices to see $W = E$, i.e., that $W^\perp \neq (0)$. We clearly have $TW \subset W$, and hence, by 3.2.4(i), we have $TW^\perp \subset W^\perp$. Then $T|W^\perp \in B(W^\perp)$ is self-adjoint, and if $W^\perp \neq (0)$, then by the preceding paragraph, there is an eigenvector for T in $W^\perp$. This clearly contradicts the maximality property of the orthonormal set generating W and proves the first assertion of 3.2.3. The remaining assertions follow from Example 3.1.4.

Theorem 3.2.3 can be easily extended to apply to a commuting family of self-adjoint compact operators.

COROLLARY 3.2.5. *Let $\{T_\alpha\}_{\alpha \in I}$ be a subset of $B(E)$ such that each T_α is compact, self-adjoint, and $T_\alpha T_\beta = T_\beta T_\alpha$ for all $\alpha, \beta \in I$. Then there is an orthonormal basis $\{e_j\}$ of E such that e_j an eigenvector for every T_α.*

We give the proof for two commuting operators S, T. The general case is similar and is left as an exercise. (Exercise 3.16)

PROOF: Let E_λ be the eigenspace for T as in Theorem 3.2.3. Then if $v \in E_\lambda$, $T(Sv) = STv = S(\lambda v)$. In other words, E_λ is an S-invariant subspace. Since $S|E_\lambda$ is self-adjoint and S (and hence $S|E_\lambda$) is compact, we may choose an orthonormal basis of E_λ consisting of eigenvectors for S (and clearly for T as well.) Taking the union over all λ gives the required orthonormal basis of E (using 3.2.4(v)).

From 3.2.5, we can also extend 3.2.3 to compact normal operators.

DEFINITION 3.2.6. *If $T \in B(E)$ (E a Hilbert space), then T is called normal if $TT^* = T^*T$.*

EXAMPLE 3.2.7: a) If $T = T^*$, clearly T is normal. If E is a complex space then $p(T)$ is normal for any polynomial p. (If E is real, then $p(T)$ is self-adjoint for any (real) polynomial.)
b)   If U is unitary, then $U^* = U^{-1}$, so U is normal. We note, however, that a unitary operator can be compact only if $\dim(E) < \infty$. (See exercise 3.1.).

REMARK 3.2.8: Given any $T \in B(E)$, we can write $T = \frac{T^*+T}{2} + i\left(\frac{T^*-T}{2i}\right)$. In other words, we have $T = T_1 + iT_2$ where T_j are self-adjoint. One easily verifies: a) This representation of T is unique; and b) T is normal if and only if T_1 and T_2 commute (Exercise 3.17). We thus have from 3.2.5 and 3.2.3:

COROLLARY 3.2.9. *If T is a compact normal operator then E has an orthonormal basis of eigenvectors of T. For each $\lambda \neq 0$, $\dim E_\lambda < \infty$, and for all $\varepsilon > 0$, $\{\lambda \mid |\lambda| \geq \varepsilon$ and $\dim E_\lambda > 0\}$ is finite.*

COROLLARY 3.2.10. *If $\{T_\alpha\}_{\alpha \in I}$ is a family of compact normal operators, and $T_\alpha T_\beta = T_\beta T_\alpha$ for all $\alpha, \beta \in I$, then E has an orthonormal basis $\{e_j\}$ such that e_j is an eigenvector for each $T_\alpha \in I$.*

REMARK 3.2.11: Let e_j be the orthonormal basis given by 3.2.9. Then $Te_j = \lambda_j e_j$ for $\lambda_j \in \mathbb{C}$, $\lambda_j \to 0$. We have $T^* e_j = \bar{\lambda}_j e_j$. Thus for $T = T_1 + iT_2$ with T_j self-adjoint, we have $T_1 e_j = \mathbf{Re}(\lambda_j)e_j$, $T_2 e_j = \mathbf{Im}(\lambda_j)e_j$.

As indicated in 3.2.7(b), the spectral theorem for compact operators is inadequate for dealing with unitary operators in infinite dimensions. We shall prove a more general spectral theorem in section 4.3 that will apply to unitary operators.

3.3 Peter-Weyl Theorem

Given any representation π of a group G, a standard approach to understanding π is to

   a)   express π in terms of irreducible representations; and
   b)   understand all the irreducible representations.

In this section we prove a basic theorem concering (a) for representations of compact groups.

DEFINITION 3.3.1. *Let G be a locally compact group with left invariant measure μ which is finite on compact subsets (cf. 2.2.1). We define the (left) regular representation of G to be the representation π of G on $L^2(G)$ given by $\big(\pi(g)f\big)(x) = f(g^{-1}x)$ (This is a continuous unitary representation by 1.3.10.)*

EXAMPLE 3.3.2: Consider the case $G = S^1$, the unit circle in $\mathbb{C}$. We have seen that an orthonormal basis of $L^2(S^1)$ is given by $\{e_n \mid n \in \mathbb{Z}\}$, where $e_n(z) = z^n$. (Example A.29) There are of course many other orthonormal bases of $L^2(S^1)$ but one of the reasons (among others) that makes $\{e_n\}$ a special choice is that it solves problem (a) above for the regular representation. More precisely, for any $a \in S^1$, $\big(\pi(a)e_n\big)(z) = e_n(az) = a^n z^n$, i.e. $\pi(a)e_n = a^n e_n$. Thus, e_n is a simultaneous eigenvector for all of the operators $\pi(a)$, $a \in S^1$. In other words

$$L^2(S^1) = \sum_{n=-\infty}^{\infty} \mathbb{C} \cdot e_n$$

and $\mathbf{C} \cdot e_n$ is a 1-dimensional (and hence irreducible) invariant subspace for π.

EXAMPLE 3.3.3: In contrast to the situation for the circle, the regular representation of $\mathbf{R}$ has no 1-dimensional invariant subspace. To see this, observe that for f in such a subspace, we have for $t \in \mathbf{R}$ that $\pi(t)f = c_t f$ for some c_t. Since π is unitary, $\|\pi(t)f\| = \|f\|$, so $|c_t| = 1$. For any interval $I \subset \mathbf{R}$,

$$\int_{I-t} |f|^2 = \int_I |\pi(t)f|^2 = \int_I |c_t f|^2 = \int_I |f|^2.$$

If $f \neq 0$, we can choose a compact I so that $\int_I |f|^2 = a > 0$. We can then clearly choose a sequence t_n such that $\{I - t_n\}$ are all mutually disjoint. Then

$$\int_{\mathbf{R}} |f|^2 \geq \sum_n \int_{I-t_n} |f|^2 = \infty$$

contradicting $f \in L^2(\mathbf{R})$. In fact, there are no finite dimensional invariant subspaces of $L^2(\mathbf{R})$ for the regular representation. For if there were, then the spectral theorem in finite dimensions (e.g. in the form of Corollary 3.2.10) would imply there is a 1-dimensional invariant subspace, which we have just seen is impossible.

The Peter-Weyl theorem describes the decomposition of the regular representation of a compact group. While more complete versions are available, we shall content ourselves here with the following form of this result.

THEOREM 3.3.4. (Peter-Weyl) *Let G be a compact group, and π the regular representation on $L^2(G)$. Then $L^2(G) = \Sigma^{\oplus} H_i$ where $\dim H_i < \infty$, and H_i is invariant and irreducible for π.*

The idea of the proof is based on the simple observation that if $S, T \in B(E)$, and $ST = TS$, then S leaves the eigenspaces of T invariant (cf. the proof of 3.2.5). Since G is compact, we will be able to construct many integral operators (compact by 3.1.5) which commute with all $\pi(g)$. The eigenspaces of a related compact self-ajoint operator will be invariant for π, and those corresponding to non-0 eigenvalues will be finite dimensional. To see how to get integral operators commuting with $\pi(g)$, we first prove the following general lemma.

LEMMA 3.3.5. *Suppose X is a compact metric space and that a topological group G acts continuously on X. Let $\mu \in M(X)$ be G-invariant and $K \in C(X \times X)$ be G-invariant; i.e., satisfies $K(gx, gy) = K(x, y)$ for all $x, y \in X, g \in G$. Let $T_K : L^2(X) \to L^2(X)$ be the associated integral operator (Example 1.2.14). Let π be the representation of G on $L^2(X)$ given by $(\pi(g)f)(x) = f(g^{-1}x)$. Then $T_K \pi(g) = \pi(g)T_K$ for all $g \in G$.*

PROOF: $\bigl(T_K \circ \pi(g)\bigr)(f)(x) = \int K(x, y)f(g^{-1}y)\, d\mu(y)$. Using invariance of μ, we may replace y by gy. Thus,

$$\bigl(T_K \circ \pi(g)\bigr)(f)(x) = \int K(x, gy)f(y)\, d\mu(y)$$
$$= \int K(g^{-1}x, y)f(y)\, d\mu(y) = \bigl[\pi(g)(T_K f)\bigr](x).$$

REMARK 3.3.6: For actions under the very general hypothesis of 3.3.5, there may well not be any G-invariant $K \in C(X \times X)$. However, if $X = G$ (so that G is a compact group), there are many. Namely, for any $\varphi \in C(G)$, let $K(x, y) = \varphi(x^{-1}y)$. Then we clearly have $K(gx, gy) = K(x, y)$ for all $x, y \in G$.

PROOF OF THEOREM 3.3.4: By Zorn's lemma, we may choose a maximal collection $\{V_j\}$ of mutually orthogonal, finite dimensional, $\pi(G)$-invariant subspaces $V_j \subset L^2(G)$ with V_j irreducible for π. Let $W = \Sigma^{\oplus}V_j$. We claim $W = L^2(G)$. If not, then we can choose $f \in W^{\perp}$, $f \neq 0$. Since $C(G)$ is dense in $L^2(G)$, we can find $\varphi \in C(G)$ such that $\langle \varphi, \bar{f} \rangle \neq 0$. Define $K \in C(G \times G)$ by $K(x, y) = \varphi(x^{-1}y)$. Then the integral operator $T = T_K$ is compact by 3.1.5 and commutes with all $\pi(g)$ by 3.3.5, 3.3.6. The operator T^*T is clearly self-adjoint and compact (by 3.1.14). Furthermore, it too commutes with all $\pi(g)$. To see this, it suffices to see that T^* commutes with all $\pi(g)$. But $\pi(g)$ is unitary, so $\pi(g) = \pi(g^{-1})^*$. Hence

$$T^*\pi(g) = T^*\pi(g^{-1})^* = (\pi(g^{-1})T)^* = (T\pi(g^{-1}))^* = \pi(g)T^*.$$

We can thus write, by the spectral theorem (3.2.3),

$$L^2(G) = \ker(T^*T) \oplus \sum_{\lambda \neq 0}^{\oplus} E_\lambda,$$

where E_λ is the corresponding eigenspace for T^*T, and for $\lambda \neq 0$, $\dim E_\lambda < \infty$. By 3.3.5, each E_λ is invariant for π.

Now let $P : L^2(G) \to W^\perp$ be the orthogonal projection. Since W is clearly G-invariant by construction, and π is unitary, $W^\perp$ will be G-invariant as well. (Namely, if $y \in W^\perp$, $g \in G$, then for all $x \in W$ we have

$$\langle \pi(g)y, x \rangle = \langle y, \pi(g)^*x \rangle = \langle y, \pi(g^{-1})x \rangle = 0$$

since W is G-invariant.) It follows easily that P also commutes with all $\pi(g)$. Therefore, $PE_\lambda \subset W^\perp$ will also be G-invariant. (Namely, $\pi(g)(PE_\lambda) = P\pi(g)E_\lambda = PE_\lambda$.) Since we clearly have $\dim P(E_\lambda) < \infty$, this will contradict the maximality of $\{V_j\}$ if we can show that at least one $PE_\lambda \neq (0)$ (for $\lambda \neq 0$.)

However, if $PE_\lambda = (0)$ for all $\lambda \neq 0$, then $\Sigma^\oplus E_\lambda \subset W$, and therefore $W^\perp \subset (\Sigma^\oplus E_\lambda)^\perp$, i.e., $W^\perp \subset \ker T^*T$. But if $T^*Tx = 0$, then $0 = \langle T^*Tx, x \rangle = \langle Tx, Tx \rangle$, so $Tx = 0$. Thus, we would have $W^\perp \subset \ker T$. However,

$$(Tf)(e) = \int K(e, y)f(y)\, dy = \int \varphi(y)f(y)\, dy = \langle \varphi, \bar{f} \rangle \neq 0.$$

By Remark 1.2.16, Tf is continuous, and hence $Tf \neq 0$ in $L^2(G)$. However, $f \in W^\perp$, and this contradicts $W^\perp \subset \ker(T)$. This completes the proof.

For specific groups, one can obtain Theorem 3.3.4 in a more explicit way, as for the circle in Example 3.3.2. We shall now indicate an alternative proof of 3.3.4 for a very natural class of groups which gives some further insight into the spaces H_i.

EXAMPLE 3.3.7: Let $G \subset GL(n, \mathbf{R})$ be a compact subgroup. We can view $GL(n, \mathbf{R})$, and hence G, as a subspace of the $n \times n$ matrices $\mathbf{R}^{n \times n}$. For each r, let P_r be the space of polynomial functions on $\mathbf{R}^{n \times n}$ of degree at most r, (i.e., polynomial functions with the variables being the matrix entries). Each $g \in G$ acts on $\mathbf{R}^{n \times n}$ simply by matrix multiplication. Furthermore, for a fixed g, this map $\mathbf{R}^{n \times n} \to \mathbf{R}^{n \times n}$ is linear, and hence if $f \in P_r$, so is $x \mapsto f(g^{-1}x)$. In particular, if we let $V_r \subset C(G)$ be given by $V_r = \{(f|G) \mid f \in P_r\}$, then $\pi(g)V_r = V_r$ where π is the regular representation. We also have $V_r \subset V_{r+1}$, $\dim V_r < \infty$, and $\bigcup V_i$ is (uniformly) dense in $C(G)$. This last assertion follows from the Stone-Weierstrass theorem (A.8). Thus, it is dense in $L^2(G)$. We now let $H_r = \{x \in V_r \mid x \perp V_{r-1}\}$. Then $\dim H_r < \infty$, $\{H_r\}$ are mutually orthogonal, $\Sigma^\oplus H_r = L^2(G)$, and H_r is invariant under $\pi(G)$. A similar argument works for $G \subset GL(n, \mathbf{C})$, (where one also needs to consider the conjugates $\overline{P}_r$.)

REMARK 3.3.8: In examples 3.3.2, 3.3.7, we see that the finite dimensional invariant subspaces of $L^2(G)$ that we constructed actually consisted of continuous functions. In fact, one can show that this is true in general. (See exercise 3.18).

Theorem 3.3.4 can be used to prove a similar result about an arbitrary unitary representation of G.

THEOREM 3.3.9. *Let σ be any continuous (strong operator topology) representation of a compact group G on a Hilbert space E. Then we can write $E = \Sigma^\oplus V_i$ where V_i is invariant and irreducible for σ, and $\dim V_i < \infty$.*

PROOF: Using Zorn's lemma exactly as in the beginning of the proof of 3.3.4, we see that it suffices to show that for any such σ we can find at least one non-trivial finite dimensional invariant subspace. Fix any $x \in E$, $\|x\| = 1$. Define the map $T : E \to C(G)$ by $(T(y))(g) = \langle y, \sigma(g)x \rangle$ for any $y \in E$. Then $|T(y)(g)| \leq \|y\|$, so T is a bounded linear map. Clearly, $T \neq 0$, since $(T(x))(e) \neq 0$. We then may view T as defining a bounded linear map $T : E \to L^2(G)$, $T \neq 0$. We claim that $T\sigma(h) = \pi(h)T$ for all $h \in G$, where

π is the regular representation. Namely,

$$
\begin{aligned}
\Big(T(\sigma(h)y)\Big)(g) &= \langle \sigma(h)y, \sigma(g)x\rangle \\
&= \langle y, \sigma(h)^*\sigma(g)x\rangle \\
&= \langle y, \sigma(h^{-1}g)x\rangle \\
&= \Big(\pi(h)\big(T(y)\big)\Big)(g).
\end{aligned}
$$

The adjoint operator $T^* : L^2(G) \to E$ (using the identification of a Hilbert space with its dual as in example 1.2.22) will satisfy $\sigma(h)^*T^* = T^*\pi(h)^*$ for all h, and since σ, π are unitary, we will have $\sigma(h)T^* = T^*\pi(h)$ for all h. Since $T \neq 0$, we have $T^* \neq 0$ (cf. 1.2.21). Thus, by Theorem 3.3.4, there is some finite dimensional $\pi(G)$-invariant subspace $V \subset L^2(G)$ such that $T^*V \neq 0$. To complete the proof, we observe that $\sigma(h)T^*V = T^*\pi(h)V = T^*V$, so that T^*V is $\sigma(G)$-invariant, and it is clearly finite dimensional.

PROBLEMS FOR CHAPTER 3

3.1 If E is a Hilbert space and $T \in B(E)$ is compact, show $T(E)$ does not contain a closed infinite dimensional subspace.

3.2 Suppose X is a finite measure space and $K_1, K_2 \in L^2(X \times X)$. Suppose T_{K_i} is the corresponding integral operator on $L^2(X)$, $i = 1, 2$. Show $T_{K_1} \circ T_{K_2}$ is also an integral operator.

3.3 Let $T \in B(L^2(X))$ and suppose T is Hilbert-Schmidt. Show there is a unique $K \in L^2(X \times X)$ such that $T = T_K$.

3.4 Suppose E is Hilbert and $T \in B(E)$ is compact. Show there is a sequence of finite rank operators T_n (i.e., $\dim T_n(E) < \infty$) such that $\|T - T_n\| \to 0$.

3.5 Suppose (X, d) is a compact metric space and $S \subset C(X)$ is a subset. Then S is called *equicontinuous* if it is "uniformly uniformly continuous". I.e., for all $\varepsilon > 0$ there is $\delta > 0$ such that $d(x, y) < \delta$ implies $|f(x) - f(y)| < \varepsilon$ for all $f \in S$. Suppose S is equicontinuous and there is a constant B such that $|f(x)| \le B$ for all $x \in X$ and $f \in S$. Show S has compact closure in $C(X)$. (This is the Arzela-Ascoli theorem.)

3.6 Let $\Omega \subset \mathbf{R}^n$ be open and $X \subset \Omega$ be compact. Show that the restriction mappings $f \mapsto (f \,|\, X)$ defines a compact operator $BC^1(\Omega) \to C(X)$. Hint: Problem 3.5.

3.7 If T is a compact linear map between Banach spaces, show T^* is also compact. Hint: problem 3.5.

3.8 a) Suppose G is a compact abelian group. Show $L^2(G) = \Sigma^\oplus V_i$ where each V_i is a one-dimensional G-invariant subspace (under the regular representation). Show $V_i = \mathbf{C}h_i$ where $h_i : G \to S^1 \subset \mathbf{C}$ is a homomorphism. b) What is this decomposition for $G = T^n (= S^1 \times \cdots \times S^1)$?

3.9 Let π be the regular representation of $\mathbf{R}^n$ on $L^2(\mathbf{R}^n)$. Suppose $V \subset L^2(\mathbf{R}^n)$ is $\pi(R^n)$-invariant. a) If $u \in L^1(\mathbf{R}^n)$ and $f \in V$, show $u * f \in V$. b) Show $V \cap C^\infty(\mathbf{R}^n)$ is dense in V.

3.10 a) Let $K : [0, 1] \times [0, 1] \to \mathbf{R}$ be the characteristic function of $\{(x, y) \mid y \ge -x + 1\}$. Find all eigenvalues of T_K acting on $L^2([0, 1])$.

b) Let E be a Hilbert space with orthonormal basis $\{e_i\}$. Let $T \in B(E)$ be strictly upper triangular with respect to this basis, i.e., $\langle Te_j, e_i \rangle = 0$ if $i \le j$. Find all eigenvalues of T.

3.11 If X is a compact metric space with a finite measure, and $K \in$

$C(X \times X)$, show the integral operator T_K defines a compact operator on $L^p(X)$ for $1 \leq p < \infty$.

3.12 Let G be the group of 2×2 matrices of the form

$$\left\{ \begin{pmatrix} x & y \\ 0 & 1 \end{pmatrix} \mid x, y \in \mathbf{R}, \ x \neq 0 \right\}.$$

Show that G is not unimodular.

3.13 Let X be a separable metric space and μ a finite measure on X. Let $\varphi \in L^\infty(X)$. Show that the multiplication operator M_φ is compact if and only if

   i) $\mu \big|_{\varphi^{-1}(\mathbf{C} - \{0\})}$ is atomic (i.e. is of the form $\sum c_i \delta_{x_i}$ where δ_{x_i} is supported on $\{x_i\}$ and $c_i > 0$;) and,

   ii) for any $\varepsilon > 0$, $\{x_i \mid |\varphi(x_i)| \geq \varepsilon\}$ is finite, where $\{x_i\}$ is as in (i).

3.14 If E is a Hilbert space with orthonormal basis $\{e_j\}$ then $T \in B(E)$ is compact if and only if $\langle Te_j, e_i \rangle \to 0$ as $i, j \to \infty$.

3.15 If $E = L^2(X)$ and $\varphi \in L^\infty(X)$, show that λ is an eigenvalue of M_φ if and only if there is a set $A \subset X$ with $\mu(A) > 0$ and $\varphi(x) = \lambda$ for all $x \in A$.

3.16 Prove the assertion of Corollary 3.2.5 in general.

3.17 Prove the assertion in Remark 3.2.8.

3.18 Suppose G is a compact group and $V \subset L^2(G)$ is finite dimensional and $\pi(G)$-invariant, where π is the regular representation. Show $V \subset C(G)$. Hint: let $\varphi \in C(G)$ and consider the integral operator constructed in 3.3.6. Show $T_K(V) \subset V$, and try to use 1.2.16.

GENERAL SPECTRAL THEORY

4.1. Spectrum of an operator

We saw in example 3.2.2 that a general bounded self-adjoint operator may have no eigenvectors at all. Thus, if we wish to obtain a theorem along the lines of the spectral theorem for compact operators in a more general setting, we need to weaken the notion of eigenvalue. In this chapter, we shall always assume $k = \mathbb{C}$.

DEFINITION 4.1.1. *Let E be a Banach space and $T \in B(E)$. Then we say that $\lambda \in \mathbb{C}$ is in the spectrum of T if $(T - \lambda I) \notin$ Aut(E), i.e., is not an invertible operator. We set $\sigma(T) = \{\lambda \in \mathbb{C} \mid \lambda$ is in the spectrum of $T\}$.*

EXAMPLE 4.1.2: a) If λ is an eigenvalue, then clearly $\lambda \in \sigma(T)$. If $\dim E < \infty$, then $\lambda \in \sigma(T)$ if and only if it is an eigenvalue.
b) Let E be a Hilbert space with orthonormal basis $\{e_i\}$. Define $T \in B(E)$ by $Te_i = \lambda_i e_i$ where $\lambda_i \to 0$ as $i \to \infty$. Suppose however that for all i, $\lambda_i \neq 0$. Then 0 is not an eigenvalue (i.e., $\ker T = 0$), but $0 \in \sigma(T)$. To see this, observe that if T is invertible we must have $T^{-1}e_i = (\lambda_i)^{-1}e_i$. Since $\lambda_i \to 0$, this cannot be a bounded operator.

EXAMPLE 4.1.3: a) Let $E = L^2(X)$, and suppose $\varphi \in L^\infty(X)$. Let T be the multiplication operator M_φ on E (example 1.2.1). Then for any $\lambda \in \mathbb{C}$, $T - \lambda I = M_{\varphi - \lambda}$. If $(T - \lambda I)^{-1}$ exists, it must be $M_{1/(\varphi - \lambda)}$. Thus, $(T - \lambda I)^{-1}$ exists if and only if $1/(\varphi(x) - \lambda) \in L^\infty(X)$. This is equivalent to $\lambda \notin$ ess range(φ) (example A.5). Thus, $\sigma(M_\varphi) = $ ess range(φ).
b) We remark that 4.1.2(b) can be considered as a special case of 4.1.3(a). Namely, if we take X to be the natural numbers with counting measure, then T in 4.1.2(b) is precisely M_φ where $\varphi(i) = \lambda_i$. (Cf. Remark 1.2.17.) The condition that $\lambda_i \to 0$ is equivalent

to the condition that ess range$(\varphi) = \{\lambda_i\} \cup \{0\}$.

c) An argument similar to that of example 3.2.2 shows that M_φ (in the generality of (a)) will have an eigenvector with eigenvalue λ if and only if there is a set $A \subset X$ with $\mu(A) > 0$ and $\varphi(x) = \lambda$ for all $x \in A$. (See exercise 3.15)

If E is a Hilbert space and $T = T^*$, we have seen that every eigenvalue is real (3.2.4(ii)). If T is unitary, then clearly every eigenvalue has modulus 1. We now wish to show that these facts are true more generally, namely for the spectrum.

DEFINITION 4.1.4. *If E is a Banach space and $T \in B(E)$, we say that T is bounded below if there is some $k > 0$ such that $\|Tx\| \geq k\|x\|$ for all $x \in E$.*

LEMMA 4.1.5. *If $T \in B(E)$, then T is invertible if and only if T is bounded below and $T(E)$ is dense.*

PROOF: If T is invertible, then we can take $k = \|T^{-1}\|^{-1}$. Conversely, if T is bijective and bounded below, then T^{-1} will be bounded and $\|T^{-1}\| \leq k^{-1}$. Thus, we need only verify that T is surjective. Given $y \in E$, choose $x_n \in E$ such that $Tx_n \to y$. Since $\{Tx_n\}$ is Cauchy and $\|Tx_n - Tx_m\| \geq k\|x_n - x_m\|$, we also have that $\{x_n\}$ is Cauchy. Since E is complete, we have $x_n \to x$ for some x. Therefore $Tx_n \to Tx$, and so $Tx = y$.

LEMMA 4.1.6. *Suppose E is a Hilbert space and $T \in B(E)$. If T, T^* are bounded below, then T is invertible.*

PROOF: By 4.1.5, it suffices to see that $T(E)$ is dense. However, from the defining equation for the adjoint, we have $T(E)^\perp = \ker T^*$, and since T^* is bounded below $T(E)^\perp = (0)$.

PROPOSITION 4.1.7. *Let E be a Hilbert space, $T \in B(E)$.*
    a) *If $T = T^*$, then $\sigma(T) \subset \mathbf{R}$.*
    b) *If T is unitary, then $\sigma(T) \subset \{z \in \mathbf{C} \mid |z| = 1\}$.*

PROOF: a) By 4.1.6, we need to show that if $|\operatorname{Im}\lambda| \neq 0$, then $T - \lambda I$ and $(T - \lambda I)^* = T - \bar{\lambda}I$ are bounded below. (It clearly suffices to do this for λ.) Let $\|x\| = 1$. Then

$$\|(T - \lambda I)(x)\| \geq |\langle (T - \lambda I)x, x \rangle| = |\langle Tx, x \rangle - \lambda\langle x, x \rangle|.$$

Since $\langle Tx, x \rangle \in \mathbb{R}$ by 3.2.4(ii) and $\|x\| = 1$, we have $\|(T - \lambda I)x\| \geq |\operatorname{Im} \lambda|$. This clearly suffices.

b) As in (a), we need to show that if $|\lambda| \neq 1$, then $T - \lambda I$ and $T^{-1} - \overline{\lambda}I$ are bounded below. (Since T^{-1} is unitary, it again suffices to check this for $T - \lambda I$.) We have

$$\|(T - \lambda I)(x)\| \geq |(\|Tx\| - \|\lambda x\|)| = |[(1 - |\lambda|)\|x\|]|$$

since $\|Tx\| = \|x\|$. That is, for all x,

$$\|(T - \lambda I)(x)\| \geq |(1 - |\lambda|)|\|x\|.$$

The following result yields the relation between the spectrum of T and that of polynomials in T.

PROPOSITION 4.1.8. *Let E be a Banach space, $T \in B(E)$ and $p \in \mathbb{C}[X]$. Then $\sigma(p(T)) = p(\sigma(T))$.*

PROOF: It suffices to prove this if p is a monic polynomial. First suppose $\lambda \in \sigma(p(T))$. Then $p(T) - \lambda I$ is not invertible. Let $\{\lambda_i\}$ be the roots of the polynomial $p(x) - \lambda$. Then we have

$$p(T) - \lambda I = \prod_{i=1}^{n} (T - \lambda_i I).$$

It follows that for some i, $T - \lambda_i I$ is not invertible, i.e., $\lambda_i \in \sigma(T)$. But $p(\lambda_i) - \lambda = 0$, so $\lambda \in p(\sigma(T))$.

Conversely, let $\lambda \in p(\sigma(T))$. I.e., for some $\alpha \in \sigma(T)$, we have $p(\alpha) - \lambda = 0$. Thus $\alpha = \lambda_{i_0}$ for some i_0, where the $\{\lambda_i\}$ are as above. Since $\lambda_{i_0} \in \sigma(T)$, $T - \lambda_{i_0}I$ is either non-injective or non-surjective (by A.14). If it is not injective, then rewriting

$$p(T) - \lambda I = (T - \lambda_1 I) \cdots (T - \lambda_n I)(T - \lambda_{i_0} I),$$

we see that $p(T) - \lambda I$ is not injective. If it is not surjective, then we write

$$p(T) - \lambda I = (T - \lambda_{i_0} I)(T - \lambda_1 I) \cdots (T - \lambda_n I)$$

and deduce that $p(T) - \lambda I$ is not surjective either. Hence $\lambda \in \sigma(p(T))$.

We now turn to the main general results on the spectrum of an operator on a Banach space. We shall need some elementary results about analytic functions with values in a Banach space. We recall first that a series $\sum_{i=1}^{\infty} x_i$ in a Banach space is called absolutely convergent if $\sum_{i=1}^{\infty} \|x_i\| < \infty$. Since

$$\|\sum_{i=n}^{m} x_i\| \leq \sum_{i=n}^{m} \|x_i\|,$$

the partial sums of an absolutely convergent series are Cauchy. Thus, absolute convergence implies convergence.

DEFINITION 4.1.9. *Let $\Omega \subset \mathbb{C}$ be an open set and F be a Banach space. A function $f : \Omega \to F$ is called analytic if for all $z_0 \in \Omega$, there is some $r > 0$ with $\{z \mid |z - z_0| < r\} \subset \Omega$ and some $x_n \in F$ such that for all $|z - z_0| < r$ we have*

$$f(z) = \sum_{n=0}^{\infty} (z - z_0)^n x_n$$

the sum converging absolutely.

The following facts are generalizations to Banach space valued functions of standard facts from the theory of functions of one complex variable. The usual proofs carry over to the case of a Banach space with no difficulty.

LEMMA 4.1.10. a) *Given a sequence $x_n \in F$, let*

$$R = \left(\overline{\lim_{n \to \infty}} \|x_n\|^{1/n} \right)^{-1}.$$

Then the series $\sum_{n=0}^{\infty} x_n (z - z_0)^n$
  i) *converges absolutely for $|z - z_0| < R$, and*
  ii) *diverges if $|z - z_0| > R$.*

b) If $f : \Omega \to F$ is analytic and $z_0 \in \Omega$, then the local expression

$$f(z) = \sum_{n=0}^{\infty} (z - z_0)^n x_n$$

is valid for $|z - z_0| < R$, where

$$R = \max\{t \mid |z - z_0| < t\} \subset \Omega.$$

c) (Liouville's theorem) If $f : \mathbb{C} \to F$ is analytic and bounded, then f is constant.

DEFINITION 4.1.11. a) If $T \in B(E)$, we let $\mathbb{C} - \sigma(T) = R(T)$. The space $R(T)$ is called the resolvent set of T.
b) Let $\|T\|_\sigma = \sup\{|\lambda| \mid \lambda \in \sigma(T)\}$. $\|T\|_\sigma$ is called the spectral radius of T.

Some basic properties of $\sigma(T)$ are contained in:

THEOREM 4.1.12. a) $\sigma(T)$ is a non-empty compact set with $\|T\|_\sigma \leq \|T\|$.
b) $R(T)$ is open and the function $f : R(T) \to B(E)$ given by $f(\lambda) = (\lambda I - T)^{-1}$ is analytic.

To prove 4.1.12, we need the following lemma.

LEMMA 4.1.13. a) If $T \in B(E)$ and $\|T\| < 1$, then $(I - T) \in \text{Aut}(E)$ and $(I - T)^{-1} = \sum_{n=0}^{\infty} T^n$.
b) If $|\lambda| > \|T\|$, then $(\lambda I - T) \in \text{Aut}(E)$ and $(\lambda I - T)^{-1} = \sum_{n=0}^{\infty} T^n / \lambda^{n+1}$.

PROOF: a) Since $\|T\| < 1$, the series $\sum T^n$ is absolutely convergent, and hence convergent, say to S. Let $S_N = \sum_{n=0}^{N-1} T^n$. Then $(I - T)S_N = S_N(I - T) = I - T^N$. Letting $N \to \infty$, we obtain $S = (I - T)^{-1}$.
b) $\lambda I - T = \lambda(I - \frac{T}{\lambda})$. Since $\|T/\lambda\| < 1$, $(\lambda I - T)^{-1}$ exists and is equal to $\lambda^{-1} \sum_{n=0}^{\infty} T^n / \lambda^n$.

PROOF OF THEOREM 4.1.12: We first observe that 4.1.13 asserts that $\|T\|_\sigma \leq \|T\|$. We next prove (b). This will prove that $\sigma(T)$ is closed and bounded. We shall deal with the assertion $\sigma(T) \neq \phi$ after proving (b).

Let $\lambda_0 \in R(T)$. Then for $\lambda \in \mathbf{C}$ we write $\lambda I - T = (\lambda - \lambda_0)I + (\lambda_0 I - T)$. Since $\lambda_0 \in R(T)$, we can write this as

$$(\lambda I - T) = (\lambda_0 I - T)(I - (\lambda_0 - \lambda)(\lambda_0 I - T)^{-1}).$$

By 4.1.13(a), we have $(\lambda I - T)$ is invertible if $\|(\lambda_0 - \lambda)(\lambda_0 I - T)^{-1}\| \leq 1$, i.e., $|\lambda - \lambda_0| < \|(\lambda_0 I - T)^{-1}\|^{-1}$. This shows that $R(T)$ is open. Furthermore, for such λ, we have (again by 4.1.13(a)) that

$$(\lambda I - T)^{-1} = (\lambda_0 I - T)^{-1} \sum_{n=0}^{\infty} (\lambda_0 - \lambda)^n (\lambda_0 I - T)^{-n},$$

i.e., $(\lambda I - T)^{-1} = \sum (\lambda - \lambda_0)^n T_n$ where $T_n = (-1)^n (\lambda_0 I - T)^{-n-1}$. This proves that $f(\lambda)$ is analytic.

Finally, we prove that $\sigma(T) \neq \emptyset$. If $\sigma(T) = \emptyset$, then $f(\lambda) = (\lambda I - T)^{-1}$ is analytic on all $\mathbf{C}$. We claim it is also bounded. By continuity, it is bounded on the compact set $\{\lambda \mid |\lambda| \leq 2\|T\|\}$. On the other hand, if $|\lambda| > 2\|T\|$, we have by Lemma 4.1.13(b) that

$$\|f(\lambda)\| \leq \sum_{n=1}^{\infty} \|T^n\| \big/ 2^{n+1} \|T\|^{n+1}.$$

Since $\|T^n\| \leq \|T\|^n$, we deduce that $\|f(\lambda)\| \leq 1/\|T\|$, showing that f is bounded. By 4.1.10(c), f is constant. This would imply $\lambda I - T$ is constant, which is clearly impossible. This completes the proof.

We now turn to obtaining more precise information on $\|T\|_\sigma$.

THEOREM 4.1.14. (Spectral radius formula) For $T \in B(E)$, $\|T\|_\sigma = \lim_{n \to \infty} \|T^n\|^{1/n}$ (and in particular, this limit exists).

PROOF: By compactness of $\sigma(T)$, $(R(T) - \{0\})^{-1} \cup \{0\}$ is an open set, and we define h on this set by

$$h(z) = \begin{cases} (z^{-1}I - T)^{-1} & \text{for } z \neq 0 \\ 0 & \text{for } z = 0 \end{cases}$$

By 4.1.12(b), h is analytic on $0 < |x| < \|T\|_\sigma^{-1}$, and by 4.1.13(b), we have for $|z| < \|T\|^{-1}$ that $h(z) = \sum_{n=0}^\infty z^{n+1} T^n$. In particular, h is analytic at 0 as well. By 4.1.10(b), we have that the expansion $h(z) = \sum z^{n+1} T^n$ holds for $|z| < \|T\|_\sigma^{-1}$, and hence that the radius of convergence, say R, for this series satisfies $R \geq \|T\|_\sigma^{-1}$. On the other hand, we cannot have $R > \|T\|_\sigma^{-1}$, for then $\lambda I - T$ would be invertible for all $|\lambda| > R^{-1}$. If $R^{-1} < \|T\|_\sigma$, this would contradict the definition of $\|T\|_\sigma$. Therefore, we have $R = \|T\|_\sigma^{-1}$, and by 4.1.10(a) we have $\|T\|_\sigma = \overline{\lim} \|T^n\|^{1/n}$.

It therefore remains only to prove that the limit exists. To see this, we remark that if $r_n \geq 0$, and $\overline{\lim} \, r_n < \infty$, to see $\lim r_n$ exists it suffices to see that for each m, $r_m \geq \overline{\lim} \, r_n$. Thus, we fix $m > 0$ and for $n > m$ write $n = qm + r$ where q, r are non-negative integers, $r < m$. Then

$$\|T^n\|^{1/n} = \|(T^m)^q T^r\|^{1/n} \leq (\|T^m\|^{1/m})^{qm/n} \|T^r\|^{1/n}$$
$$\leq (\|T^m\|^{1/m})^{1-r/n} \|T\|^{r/n}.$$

As $n \to \infty, r/n \to 0$. Thus, $\overline{\lim} \|T^n\|^{1/n} \leq \|T^m\|^{1/m}$, and this completes the proof.

We have the following important consequence.

COROLLARY 4.1.15. *If E is a Hilbert space and $T \in B(E)$ is normal, then $\|T\|_\sigma = \|T\|$.*

PROOF: By 4.1.14, it suffices to see that $\|T^n\| = \|T\|^n$ for any infinite set of positive integers. In particular, it suffices to see this for powers of 2, and by induction to see that $\|T^2\| = \|T\|^2$. We have

$$\|T\|^2 = \sup_{\|x\| \leq 1} |\langle Tx, Tx \rangle| = \sup_{\|x\| \leq 1} |\langle T^*Tx, x \rangle|.$$

By 3.2.4(iv), (since T^*T is self-adjoint), we thus have

$$\|T\|^2 = \|T^*T\|.$$

However, we also have

$$\|T^*T\| = \sup_{\|x\|=1} |\langle T^*Tx, T^*Tx \rangle|^{1/2}$$
$$= \sup_{\|x\|=1} |\langle T^2 x, T^2 x \rangle|^{1/2} \qquad \text{(since } TT^* = T^*T\text{)}$$
$$= \|T^2\|.$$

EXAMPLE 4.1.16: We consider some examples for 2×2 matrices.
a) $T = \begin{pmatrix} \lambda_1 & 0 \\ 0 & \lambda_2 \end{pmatrix}$ is normal for $\lambda_i \in \mathbf{C}$. We have $\sigma(T) = \{\lambda_1, \lambda_2\}$,
$\|T\|_\sigma = \max\{|\lambda_1|, |\lambda_2|\}$. Clearly $\|T\| = \|T\|_\sigma$. We have

$$T^n = \begin{pmatrix} \lambda_1^n & 0 \\ 0 & \lambda_2^n \end{pmatrix}$$

and $\|T^n\|^{1/n} = \|T\|$.

b) Let $T = \begin{pmatrix} 1 & \lambda \\ 0 & 1 \end{pmatrix}$, $\lambda \neq 0$. Then T is not normal. We have
$\sigma(T) = \{1\}$, $\|T\|_\sigma = 1$, but $\|T\| = (1 + |\lambda|^2)^{1/2}$. Then $T^n = \begin{pmatrix} 1 & n\lambda \\ 0 & 1 \end{pmatrix}$, $\|T^n\| = (1 + |n\lambda|^2)^{1/2}$, and we see $\|T^n\|^{1/n} \to 1$, as required by 4.1.14.

An examination of the proofs of 4.1.12–4.1.14 shows that only certain properties of $B(E)$ were used. Namely, we make the following definition.

DEFINITION 4.1.17. *A Banach algebra B is a Banach space which is also an algebra such that $\|xy\| \leq \|x\|\,\|y\|$ for all $x, y \in B$. If B has an identity e, we also require that $\|e\| = 1$.*

As basic examples, we have $B(E)$ where E is a Banach space, and $C(X)$ where X is compact and multiplication is simply pointwise multiplication. We shall consider a few more examples in section 4.2, but here we just record the fact that 4.1.12–4.1.14 hold for Banach algebras with identity, with precisely the same proofs. To state these results we need:

DEFINITION 4.1.18. *Let B be a Banach algebra with identity. For $x \in B$, let $\sigma(x) = \{\lambda \in \mathbf{C} \mid x - \lambda e$ does not have an inverse in $B\}$. Let $R(x) = \mathbf{C} - \sigma(x)$. We define $\|x\|_\sigma = \sup\{|\lambda| \mid \lambda \in \sigma(x)\}$.*

THEOREM 4.1.19. *Let B be a Banach algebra with identity, and let $x \in B$. Then:*
a) *$\sigma(x)$ is a non-empty compact set, and $\|x\|_\sigma \leq \|x\|$.*
b) *$R(x)$ is open and the function $f : R(x) \to B$ given by $f(x) = (\lambda e - x)^{-1}$ is analytic.*
c) *$\|x\|_\sigma = \lim_{n \to \infty} \|x^n\|^{1/n}$ (and in particular, this limit exists.)*
d) *If $\|x\| < 1$, then $e - x$ is invertible.*

As indicated above, the proof of 4.1.19 is identical to that of 4.1.12–4.1.14.

4.2. The spectral theorem for self-adjoint operators

The spectral theorem for compact self-adjoint operators gives us an essentially complete description of such operators. We now generalize that result to the non-compact case. For convenience, we shall assume all our Hilbert spaces are separable.

DEFINITION 4.2.1. *Let E_1, E_2 be Hilbert spaces, $T_i \in B(E_i)$. Then T_1 and T_2 are called unitarily equivalent if there is a unitary map $U : E_1 \to E_2$ such that $U T_1 U^{-1} = T_2$.*

EXAMPLE 4.2.2: Let $E_1 = \ell^2(\mathbf{Z}^+)$. Let $\lambda_i \in \mathbf{C}$ with $|\lambda_i| \to 0$, and let $T_1 \in B(E)$ be given by $T_1 x_i = \lambda_i x_i$, where $\{x_i\}$ is the orthonormal basis $x_i = \chi_{\{i\}}$, the characteristic function of $\{i\}$. If $T_2 \in B(E_2)$ is any compact self-adjoint operator, then the spectral theorem for compact operators (3.2.3) asserts that T_2 is unitarily equivalent to an operator of the form T_1, for a suitable choice of $\{\lambda_i\}$, namely $\{\lambda_i\}$ is the set of eigenvalues of T_2. Thus, the spectral theorem for compact operators can be expressed as saying that every compact self-adjoint operator is unitarily equivalent to a "model example" of such an operator. We recall (cf. 1.2.17) that T_1 can also be viewed as a multiplication operator M_φ, $\varphi \in \ell^\infty(\mathbf{Z}^+)$, on $\ell^2(\mathbf{Z}^+)$, where $\varphi(i) = \lambda_i$. Thus, Theorem 3.2.3 can be viewed as the assertion that every compact self-adjoint operator is unitarily equivalent to a multiplication operator on a discrete measure space. Here is one form of the generalization we shall prove. We shall see a more precise statement later.

THEOREM 4.2.3. *(Spectral theorem for self-adjoint operators) Let $T \in B(E)$ be self-adjoint. Then T is unitarily equivalent to a*

multiplication operator. More precisely, there is a measure space
(X, μ) *and* $\varphi \in L^\infty(X)$ *such that* T *and* M_φ *are unitarily equivalent*
(where M_φ *acts on* $L^2(X)$*).*

This result has the effect of reducing questions about T to
questions about M_φ, where the situation is often transparent.

In proving 4.2.3, it will be important to consider not just T
but $p(T)$ for any polynomial p. The set $\{p(T)\} \subset B(E)$ is of course
a subalgebra. It will be useful to have some properties of this and
related algebras, and hence we make the following definition.

DEFINITION 4.2.4. *A* C^**-algebra is a Banach algebra* $\mathcal{A}$ *with an*
additional operation $\mathcal{A} \to \mathcal{A}$, $x \mapsto x^*$, *such that*

  i) $x^{**} = x$
  ii) $*$ *is conjugate linear, i.e.,* $(cx)^* = \bar{c}x^*$, $(x+y)^* = x^* + y^*$.
  iii) $(xy)^* = y^* x^*$
  iv) $\|x^*x\| = \|x\|^2$.
  v) *If* $\mathcal{A}$ *has an identity* I, *then* $I^* = I$.

EXAMPLE 4.2.5: a) If X is a compact space, then $C(X)$ is a C^*-
algebra with identity, under pointwise multiplication, with $f^* = \bar{f}$.
This C^*-algebra is clearly commutative.
b) If (X, μ) is a measure space, then $L^\infty(X)$ is a commutative
C^*-algebra with identity, again with pointwise multiplication and
$f^* = \bar{f}$.

EXAMPLE 4.2.6: Let E be a Banach space. Then $B(E)$ is a Banach algebra with identity under composition. If E is a Hilbert
space, then $B(E)$ is a C^*-algebra where T^* is the adjoint. (To see
(iv), observe that $\|T^*T\| = \sup_{\|x\| \leq 1} |\langle T^*Tx, x \rangle|$ by 3.2.4(iv), since
T^*T is self-adjoint. Thus, $\|T^*T\| = \sup_{\|x\| \leq 1} |\langle Tx, Tx \rangle| = \|T\|^2$.)
Clearly $B(E)$ is a non-commutative C^*-algebra.

EXAMPLE 4.2.7: a) If $\mathcal{A}$ is a C^*-algebra any closed subalgebra
$B \subset \mathcal{A}$ which contains I and is closed under $*$ is a C^*-algebra. If
B is not closed but is a subalgebra such that $I \in B$ and $B^* = B$,
then $\overline{B} \subset \mathcal{A}$ will be a C^*-algebra.
b) For example, let $T \in B(E)$. Then $\{p(T) \mid p$ is a polynomial$\}$ is
clearly a subalgebra of $B(E)$, but it will not in general be closed
under $*$. However, if $T = T^*$, then $\overline{\{p(T)\}}$ will be a commutative

C^*-subalgebra of $B(E)$. If T is normal, then $\{p(T,T^*) \mid p$ is a polynomial in 2 variables$\}$ will be a commutative subalgebra closed under $*$, and hence its closure will be a commutative C^*-subalgebra of $B(E)$.

DEFINITION 4.2.8. *If $\mathcal{A}_1, \mathcal{A}_2$ are C^*-algebras, a map $M : \mathcal{A}_1 \to \mathcal{A}_2$ is called an isomorphism if M is an isometric isomorphism of Banach spaces, M is a homomorphism of algebras with identity, and $M(x^*) = M(x)^*$. (That is, as one expects, M preserves all the structures present.)*

EXAMPLE 4.2.9: Let (X, μ) be a measure space. Then $M : L^\infty(X) \to B(L^2(X))$, $M(\varphi) = M_\varphi$ (the multiplication operator) is an isomorphism of the C^*-algebra $L^\infty(X)$ with its image, i.e., with a commutative C^*-algebra of operators on $L^2(X)$. It is isometric by 1.2.1 (cf. 1.2.3), is clearly an algebra homomorphism, and $M_{\bar\varphi} = (M_\varphi)^*$.

In example 4.2.9, functions are converted into operators. The problem we confront in proving Theorem 4.2.3 is the converse , namely to convert operators into functions. The following result, which is really just a reorganization of some of the results of section 4.1, does this, but not in the form asserted in 4.2.3. It is, however, a basic step on the way to 4.2.3. For a polynomial $p \in \mathbb{C}[X]$, and subset $A \subset \mathbb{C}$, we regard $p|A$ as an element of $C(A)$. (Of course, $p|A$ may be identically 0.) We let $P(A) \subset C(A)$ be given by $P(A) = \{(p|A) \mid p \in \mathbb{C}[X]\}$.

THEOREM 4.2.10. *Let $T \in B(E)$ be self-adjoint, and $\mathcal{A}_T \subset B(E)$ be the closure of $\{p(T) \mid p \in \mathbb{C}[X]\}$. (Thus, $\mathcal{A}_T$ is a commutative C^*-subalgebra of $B(E)$; cf. 4.2.7.) Then there is a unique isomorphism of C^*-algebras $\Phi : C(\sigma(T)) \to \mathcal{A}_T$ such that $\Phi(p|\sigma(T)) = p(T)$ for all $p \in \mathbb{C}[X]$.*

PROOF: The map $\psi : \mathbb{C}[X] \to \mathcal{A}_T$ given by $p \mapsto p(T)$ is clearly an algebra homomorphism. Furthermore, $\bar{p}(T) = p(T)^*$, where $\bar{p}$ is obtained by taking complex conjugation of all coefficients. We also have the algebra homomorphism $\mathbb{C}[X] \to P(\sigma(T))$, given by $p \mapsto p|\sigma(T)$. If $p|\sigma(T) = 0$, then by Proposition 4.1.8, $\sigma(p(T)) = \{0\}$, i.e., $\|p(T)\|_\sigma = 0$. Since $p(T)$ is normal, Corollary 4.1.15

implies $p(T) = 0$. It follows that the map ψ factors to a map $\Phi : P(\sigma(T)) \to \mathcal{A}_T$ such that $\Phi(p|\sigma(T)) = p(T)$. To ease notation we shall now simply write $p|\sigma(T)$ as p. As an element of $C(\sigma(T))$, we have $\|p\| = \sup\{p(\lambda) \mid \lambda \in \sigma(T)\}$, which by 4.1.8 gives $\|p\| = \|p(T)\|_\sigma$. Once again by 4.1.15 then we have $\|p\| = \|p(T)\|$. That is, Φ is an isometry into the Banach space $\mathcal{A}_T$. By Stone-Weierstrass [A.8], $P(\sigma(T)) \subset C(\sigma(T))$ is dense, and hence Φ extends uniquely to an isometry $\Phi : C(\sigma(T)) \to \mathcal{A}_T$. Since $\Phi(P(\sigma(T)))$ is dense in $\mathcal{A}_T$ by definition, $\Phi : C(\sigma(T)) \to \mathcal{A}_T$ is an isometric isomorphism. That Φ is an algebra homomorphism and that it commutes with * follow easily from the fact that these are true on $P(\sigma(T))$.

Via the isomorphism of Theorem 4.2.10 we have associated to each operator in $\mathcal{A}_T$ (in particular to T) a continuous function on $\sigma(T)$. However, in this association we have not kept track of the Hilbert space E. Injecting this further information into the framework of Theorem 4.2.10 will enable us to prove 4.2.3. We first need a definition.

DEFINITION 4.2.11. a) If $T \in B(E)$, a vector $x \in E$ is called T-cyclic if the smallest closed T-invariant subspace of E containing x is E itself. Equivalently, $\{p(T)x \mid p$ a polynomial$\}$ is dense in E. b) More generally, if $B \subset B(E)$ is a subalgebra we say that x is B-cyclic if Bx is dense in E.

The study of operators or algebras of operators is facilitated by:

PROPOSITION 4.2.12. Let $\mathcal{A} \subset B(E)$ be a subalgebra which is closed under *. Then $E = \Sigma^{\oplus} E_i$, where $\{E_i\}$ are mutually orthogonal $\mathcal{A}$-invariant subspaces, each possessing a cyclic vector.

PROOF: By Zorn's lemma, we let $V \subset E$ be a closed $\mathcal{A}$-invariant subspace maximal with respect to the existence of such a decompostion $V = \Sigma^{\oplus} V_i$ of V. By Lemma 3.2.4(i), $V^{\perp}$ is also $\mathcal{A}$-invariant. If $V^{\perp} \neq (0)$, choose any $x \in V^{\perp}$, $x \neq 0$. Then the decomposition $V \oplus \overline{\mathcal{A}x} = \overline{\mathcal{A}x} \oplus \Sigma^{\oplus} V_i$ contradicts maximality of V. Thus, $V = E$.

The following result, combined with Theorem 4.2.10 proves the spectral theorem (4.2.3) for operators with a cyclic vector.

THEOREM 4.2.13. *Let $\mathcal{A} \subset B(E)$ be a commutative C^*-subalgebra with $I \in \mathcal{A}$, and suppose there is an $\mathcal{A}$-cyclic vector in E. Suppose further that there is an isomorphism of C^*-algebras $F : \mathcal{A} \to C(X)$, where X is a compact metrizable space. Then there is a measure μ on X such that all elements of $\mathcal{A}$ are simultaneously unitary equivalent to multiplication operators on $L^2(X)$. More precisely, there is a unitary map $U : L^2(X) \to E$ such that $U^{-1}TU = M_{F(T)}$ for all $T \in \mathcal{A}$.*

PROOF: Let $v \in E$ be an $\mathcal{A}$-cyclic vector. We can assume $\|v\| = 1$. For $f \in C(X)$, denote $F^{-1}(f)$ by T_f. Define $\mu(f) = \langle T_f v, v \rangle$. Then $\mu : C(X) \to \mathbb{C}$ is linear, and $\|\mu(f)\| \leq \|T_f\| = \|f\|$ (since F is an isometry). That is, $\mu \in C(X)^*$, $\|\mu\| \leq 1$, and in fact $\|\mu\| = 1$ since if 1 is the constant function with value 1, then $\mu(1) = \langle T_1 v, v \rangle = \langle Iv, v \rangle = 1$. Suppose $f \in C(X)$ is non-negative. Then $f = \bar{g}g$ for some $g \in C(X)$, and hence $\mu(f) = \langle T_{\bar{g}g}v, v \rangle = \langle T_g v, T_g v \rangle \geq 0$ since F is an isomorphism of C^*-algebras. Thus, by the Riesz representation theorem (A.19), μ defines a probability measure on X, which we continue to denote by μ. That is, we have $\mu \in M(X)$ such that for all $f \in C(X)$, $\int f \, d\mu = \langle T_f v, v \rangle$.

Now define $U : C(X) \to E$ by $Uf = T_f v$. We claim this is an isometry for $C(X)$ considered as a subspace of $L^2(X)$, i.e., for $C(X)$ with the L^2-norm. Namely,

$$\begin{aligned}
\|Uf\|^2 = \|T_f v\|^2 = \langle T_f v, T_f v \rangle &= \langle T_f^* T_f v, v \rangle \\
&= \langle T_{\bar{f}f} v, v \rangle \\
&= \int \bar{f} f \, d\mu \\
&= \|f\|_2^2.
\end{aligned}$$

Furthermore, $U(C(X)) = \mathcal{A}v$ is dense in E. Thus, U extends to a unitary isomorphism $L^2(X) \to E$.

Finally, we see that for $f, g \in C(X)$ that

$$\left(U^{-1}T_f U\right)g = U^{-1}T_f T_g v = U^{-1}T_{fg}v = fg = M_f g.$$

Since $C(X) \subset L^2(X)$ is dense, we have $U^{-1}T_f U = M_f$ for all $f \in C(X)$, and letting $f = F(T)$ we obtain the theorem.

COROLLARY 4.2.14. (Spectral theorem for self-adjoint operators)
   a) (Theorem 4.2.3) *Every self-adjoint operator is unitary
      equivalent to a multiplication operator.*
   b) *If T has a cyclic vector, then up to unitary equivalence we
      may take $E = L^2(\sigma(T))$ for some probability measure on
      $\sigma(T)$, and T to be multiplication by the function $\varphi(x) =
      x$.*

PROOF: b) follows immediately from Theorems 4.2.10 and 4.2.13.
To see a), by 4.2.12 we can write $E = \Sigma^{\oplus}E_i$ where $T_i = T|\,E_i$
has a cyclic vector. Applying (b), we can find $\mu_i \in M(\sigma(T_i))$ and
a unitary operator $U_i : E_i \to L^2(\sigma(T_i), \mu_i)$ such that $U_i T_i U_i^{-1} =
M_{\varphi_i}$ where $\varphi_i \in L^{\infty}(\sigma(T_i))$ is given by $\varphi_i(x) = x$. Define

$$U = \Sigma^{\oplus}U_i \; : \; E \to \sum_i^{\oplus} L^2(\sigma(T_i), \mu_i)$$

Let (X, μ) be the disjoint union of the measure spaces $(\sigma(T_i), \mu_i)$.
Define $\varphi : X \to \mathbf{R}$ by $\varphi|X_i = \varphi_i$. Since $\sigma(T_i) \subset \sigma(T)$, we have
$\|\varphi_i\|_{\infty} \leq \|T\|_{\sigma}$ for all i, and hence $\varphi \in L^{\infty}(X)$. We have a natural
unitary isomorphism $V : \Sigma^{\oplus}L^2(\sigma(T_i), \mu_i) \to L^2(X, \mu)$ such that

$$V \circ (\Sigma^{\oplus} M_{\varphi_i}) \circ V^{-1} = M_{\varphi}.$$

Then $V \circ U : E \to L^2(X)$ gives a unitary equivalence of T with
M_{φ}.

REMARK 4.2.15: a) One can easily strengthen the assertion of
4.2.14(a) to assert that the measure in question be finite. See exercise 4.4.
b). Assertion (b) is of course much more explicit than (a). One
can generalize the formulation in (b) to obtain the same type of
statement for a general self-adjoint operator without assuming the
existence of a cyclic vector. We now sketch this development, leaving the details as an exercise. First suppose (X, μ) is a measure
space and H is a Hilbert space. Then

$$L^2(X; H) = \{f : X \to H \mid \int \|f(x)\|_H^2 \, d\mu(x) < \infty\}$$

is a Hilbert space with inner product

$$\langle f, g \rangle = \int \langle f(x), g(x) \rangle_H \, du(x).$$

More generally, for each $i \in \{\infty\} \cup \mathbf{Z}^+$, let H_i be a (separable) Hilbert space of dimension i. Let $\{X_i\}_{i \in \infty \cup \mathbf{Z}^+}$ be a partition of X, i.e., a disjoint decomposition of X into measurable sets (some of which may be null). Let $\mathcal{P}$ be this partition (together with its indexing). If we have an assignment $x \mapsto f(x)$ where $f(x) \in H_i$ for $x \in X_i$, we say that f is measurable if $f \, | X_i$ is measurable for all i. We let

$$L^2(X; \mathcal{P}) = \{ f \, | \, f(x) \in H_i \text{ for } x \in X_i,$$

and such that
$$\int_X \|f(x)\|^2 \, d\mu(x) < \infty \}.$$

Then $L^2(X; \mathcal{P})$ is a Hilbert space. If the partition is trivial, i.e., $X = X_n$ for some fixed n, then $L^2(X; \mathcal{P}) = L^2(X; H_n)$. In particular, if $X = X_1$, then $L^2(X; \mathcal{P}) = L^2(X)$.

In general, we have a natural unitary isomorphism

$$L^2(X; \mathcal{P}) \cong \Sigma^{\oplus}_{i \in \{\infty\} \cup \mathbf{Z}^+} L^2(X_i; H_i).$$

For any $\varphi \in L^\infty(X)$, we then have a generalized multiplication operator M_φ on $L^2(X; \mathcal{P})$ given by $(M_\varphi f)(x) = \varphi(x)f(x)$. One easily verifies that $\|M_\varphi\| = \|\varphi\|_\infty$, as in the usual case.

We can then formulate the spectral theorem in general as:

COROLLARY 4.2.16. (Spectral theorem) *Let E be a Hilbert space, and $T \in B(E)$ be self-adjoint. Then there is a measure $\mu \in M(\sigma(T))$ and an indexed partition $\mathcal{P}$ of $(\sigma(T), \mu)$ such that T is unitarily equivalent to the multiplication operator M_φ on $L^2(\sigma(T); \mathcal{P})$, where $\varphi(x) = x$.*

PROOF: We first indicate the proof for the case where $E = E_1 \oplus E_2$, where each E_i is T-invariant and has a cyclic vector. As in the proof of 4.2.14(a), we have T is unitary equivalent to $M_{\varphi_1} \oplus M_{\varphi_2}$ on

$$L^2\big(\sigma(T), \mu_1\big) \oplus L^2\big(\sigma(T), \mu_2\big)$$

where $\varphi_i(x) = x$. We can write $\sigma(T)$ as a disjoint union, $\sigma(T) = A_1 \cup A_2 \cup A_{12}$ where $\mu_i(A_{3-i}) = 0$ and $\mu_1|A_{12} \sim \mu_2|A_{12}$. Then M_{φ_i} is unitary equivalent to

$$M_{\varphi_i|A_i} \oplus M_{\varphi_i|A_{12}}$$

on

$$L^2\big(\sigma(T), \mu_i|A_i\big) \oplus L^2\big(\sigma(T), \mu_i|A_{12}\big).$$

Thus, T is unitarily equivalent to

$$M_{\varphi_1|A_1} \oplus M_{\varphi_2|A_2} \oplus M_{\varphi_1|A_{12}} \oplus M_{\varphi_2|A_{12}}$$

on

$$L^2\big(\sigma(T), \mu_1|A_1\big) \oplus L^2\big(\sigma(T), \mu_2|A_2\big) \oplus$$
$$L^2\big(\sigma(T), \mu_1|A_{12}\big) \oplus L^2\big(\sigma(T), \mu_2|A_{12}\big).$$

Since $\mu_1|A_{12} \sim \mu_2|A_{12}$, and $\varphi_1 = \varphi_2$, we have $M_{\varphi_1|A_{12}} \oplus M_{\varphi_2|A_{12}}$ is unitary equivalent to M_φ on $L^2((\sigma(T), \mu_1|A_{12}); \mathbf{C}^2)$. (Cf. exercise 4.4.) Thus, letting $X_1 = A_1 \cup A_2$, $X_2 = A_{12}$, we have that T is unitarily equivalent to M_φ on $L^2\big((\sigma(T), \mu); \mathcal{P}\big)$ where $\mathcal{P} = \{X_1, X_2\}$ and $\mu \sim \mu_1 + \mu_2$. The argument in general (i.e., for an arbitrary decomposition into T-cyclic subspaces) is similar, but with the usual notational (and some measure theoretic) complications. We leave this as an exercise.

As in the case of compact operators, one would like a spectral theorem for a commuting family of self-adjoint operators, which would in particular yield a spectral theorem for normal (and hence unitary) operators. Theorem 4.2.13 gives us the required conclusion. The hypotheses of 4.2.13 are satisfied by the C^*-algebra generated by a single self-adjoint operator by Theorem 4.2.10. In the next section, we see how to obtain the hypotheses of 4.2.13 more generally.

4.3. Gelfand's theory of commutative C^*-algebras

The aim of this section is to show that for any commutative C^*-algebra of operators $\mathcal{A} \subset B(E)$ that there is a compact space X such that $\mathcal{A}$ is isomorphic to $C(X)$. (Furthermore, if $\mathcal{A}$ is separable, so is X.) As a corollary of 4.2.13, we then deduce that the elements of $\mathcal{A}$ are simultaneously unitarily equivalent to multiplication operators, exactly as in the case of a single self-adjoint transformation.

Given any Banach space B, we can always realize B as a space of functions, namely as functions on B^*. Thus, for each $x \in B$, we have $e_x : B^* \to \mathbb{C}$ given by $e_x(\lambda) = \lambda(x)$. By restricting this to B_1^* we obtain a map $B \to C(B_1^*)$, where B_1^* is a compact space with the weak $*$-topology. The problem with this for our purposes is that even if B is an algebra, the map $B \to C(B_1^*)$ is not in general an algebra homomorphism. To remedy this, we make the following definition.

DEFINITION 4.3.1. *Let B be a Banach algebra with identity. We let $\widehat{B} = \{\lambda \in B^* \mid \lambda : B \to \mathbb{C}$ is an algebra homomorphism, with $\lambda(1) = 1\}$.*

It is straightforward that $\widehat{B}$ is closed in B^* with the weak $*$-topology. Thus $\widehat{B}_1 = B_1^* \cap \widehat{B}$ is a compact subset of B_1^* with the weak $*$-topology. Since for $\lambda \in \widehat{B}$ we have

$$e_{xy}(\lambda) = \lambda(xy) = \lambda(x)\lambda(y) = e_x(\lambda)e_y(\lambda),$$

the map $x \mapsto e_x$ defines an algebra homomorphism.

DEFINITION 4.3.2. *If B is a Banach algebra with identity, the algebra homomorphism $\Phi : B \to C(\widehat{B}_1)$ given by $\big(\Phi(x)\big)(\lambda) = \lambda(x)$ is called the Gelfand transform of B.*

While we have dealt with one problem by introducing $\widehat{B}$, we have created another, namely the question of the existence of elements in $\widehat{B}$. At this point, we have not even established that $\widehat{B} \neq \emptyset$. However, the main result of this section is:

THEOREM 4.3.3. *If E is a Hilbert space and $\mathcal{A} \subset B(E)$ is a commutative C^*-algebra of operators with $I \in \mathcal{A}$ then $\widehat{\mathcal{A}_1} = \widehat{\mathcal{A}}$ and the Gelfand transform is an isomorphism of C^*-algebras.*

Before turning to the proof of 4.3.3, we give some examples.

EXAMPLE 4.3.4: Let X be a compact (Hausdorff) space. For each $x \in X$, let $\lambda_x \in C(X)^*$ be $\lambda_x(f) = f(x)$. (That is, λ_x corresponds to the point measure at x.) Clearly, $\lambda_x \in \widehat{C(X)}$. We claim in fact that $X \to \widehat{C(X)}$, $x \mapsto \lambda_x$, is a homeomorphism. It is clearly injective, continuous, and since X is compact, it suffices to see it is surjective. Let $\lambda \in \widehat{C(X)}$. We claim first that there is some $x \in X$ such that $f(x) = 0$ whenever $\lambda(f) = 0$. This suffices, for then we have for each f that $\lambda(f - \lambda(f) \cdot 1) = 0$; therefore $f(x) - \lambda(f) = 0$, and hence $\lambda = \lambda_x$. If no such x exists then for each x choose $f_x \in C(X)$ such that $\lambda(f_x) = 0$ but $f_x(x) \neq 0$. By replacing f_x by either $\mathbf{Re}(f_x)$ or $\mathbf{Im}(f_x)$, we can assume f_x is real valued, and by taking f_x^2 (and using the fact that λ is multiplicative), we can assume $f_x(y) \geq 0$ for all $y \in X$ and $f_x(x) > 0$. By compactness of x, we can choose a finite set $x_1 \ldots, x_n \in X$ such that $g = \sum f_{x_i}$ is strictly positive on X. Since $\lambda(g \cdot \frac{1}{g}) = \lambda(g)\lambda(\frac{1}{g})$, and $\lambda(1) = 1$, we have $\lambda(g) \neq 0$, contradicting the assumption that $\lambda(f_{x_i}) = 0$.

EXAMPLE 4.3.5: Let $T \in B(E)$ be self-adjoint. Then by 4.2.10, $\mathcal{A}_T \cong C(\sigma(T))$. Thus, by 4.3.4, we have a homeomorphism $\widehat{\mathcal{A}_T} \cong \sigma(T)$. Combining the explicit isomorphisms of 4.2.10 and 4.3.4 we can state this as follows. For each $\alpha \in \sigma(T)$, there exists a unique $\lambda_\alpha \in \widehat{\mathcal{A}_T}$ such that $\lambda_\alpha(T) = \alpha$, and $\lambda_\alpha(p(T)) = p(\alpha)$ for any polynomial p. Every $\lambda \in \widehat{\mathcal{A}_T}$ is of the form λ_α for some (unique) $\alpha \in \sigma(T)$.

From Example 4.3.5 we deduce two facts about $\widehat{\mathcal{A}}$ for a C^*-algebra of operators.

LEMMA 4.3.6. *Let $\mathcal{A} \subset B(E)$ be a commutative C^*-algebra of operators, $I \in \mathcal{A}$. Then:*
   i) *For all $\lambda \in \widehat{\mathcal{A}}$ and $T \in \mathcal{A}$, $\lambda(T^*) = \overline{\lambda(T)}$. Hence, the Gelfand transform commutes with $*$.*
   ii) *For $\lambda \in \widehat{\mathcal{A}}$ we have $\|\lambda\| = 1$. In particular, $\widehat{\mathcal{A}} = \widehat{\mathcal{A}_1}$.*

PROOF: i) Since $\lambda \mid \mathcal{A}_T \in \widehat{\mathcal{A}}_T$ for any $T \in \mathcal{A}$, we deduce from 4.3.5 that $\lambda(T) \in \mathbf{R}$ if $T = T^*$. (Recall $\sigma(T) \subset \mathbf{R}$ if $T = T^*$.) If $S \in \mathcal{A}$, we can write $S = T_1 + iT_2$ where $T_j^* = T_j$ (cf. 3.2.8), and $S^* = T_1 - iT_2$. Thus

$$\lambda(S^*) = \lambda(T_1) - i\lambda(T_2) = \overline{\lambda(S)}.$$

ii) From 4.3.5, we see that if $T \in \mathcal{A}$ and $T = T^*$, $|\lambda(T)| \leq \|T\|_\sigma = |T|$ for all $\lambda \in \widehat{\mathcal{A}}_T$, and in particular for all $\lambda \in \widehat{\mathcal{A}}$. Given any $S \in \mathcal{A}$ with $\|S\| \leq 1$, we have that S^*S is self-adjoint and $\|S^*S\| \leq 1$. Thus, $|\lambda(S^*S)| \leq 1$. However, by (i),

$$\lambda(S^*S) = \overline{\lambda(S)}\lambda(S) = |\lambda(S)|^2.$$

Thus, $|\lambda(S)| \leq 1$, so $\|\lambda\| \leq 1$. Since $\lambda(1) = 1, \|\lambda\| = 1$.

As we discussed above, a main problem for a general $\mathcal{A}$ is the existence of elements in $\widehat{\mathcal{A}}$. For $\mathcal{A}^*$, of course, the Hahn-Banach theorem assures the existence of many elements. For $\widehat{\mathcal{A}}$, we then have the following analogous basic result. We first observe that if B is a Banach algebra with identity and $\lambda \in \widehat{B}$, then any $x \in B$ for which $\lambda(x) = 0$ cannot be invertible. (This follows from the fact that $\lambda(x)\lambda(x^{-1}) = \lambda(xx^{-1}) = \lambda(1) = 1$.) The presence of sufficiently many elements of $\widehat{B}$ is then given by:

THEOREM 4.3.7. *Suppose B is a commutative Banach algebra with identity. If $x \in B$ is not invertible, then there is some $\lambda \in \widehat{B}$ such that $\lambda(x) = 0$.*

The key to the proof of 4.3.7 is the following observation.

LEMMA 4.3.8. (Gelfand-Mazur) *If B is a Banach algebra with identity and B is a division ring (i.e., all non-zero elements are invertible) then $B \cong \mathbf{C}$, i.e. B consists of scalar multiples of e.*

PROOF: If $x \in B$, there is some $\lambda \in \mathbf{C}$ such that $x - \lambda e$ is not invertible (Theorem 4.1.19). Then $x - \lambda e = 0$, i.e., $x \in \mathbf{C} \cdot e$.

PROOF OF 4.3.7: As in any commutative ring with identity, for any non-invertible $x \in B$ there is, by Zorn's lemma, a maximal (proper) ideal $I \subset B$ with $x \in I$. We claim that I is closed. Clearly its closure $\overline{I}$ is also an ideal, so it suffices to see, by maximality, that $\overline{I}$ is also proper. However, if I is proper, then a neighborhood of e fails to intersect I by 4.1.19, and hence $e \notin \overline{I}$. Since I is a maximal ideal, B/I is a field and since I is closed, B/I is a Banach field (Exercise 1.9). By 4.3.8, $B/I \cong \mathbb{C}$. Thus for $\lambda : B \to B/I \cong \mathbb{C}$, we have $\lambda \in \widehat{B}$ and $\lambda(x) = 0$.

From 4.3.7, we have:

COROLLARY 4.3.9. *Let E be a Hilbert space and $\mathcal{A} \subset B(E)$ a commutative C^*-algebra with $I \in \mathcal{A}$. Let $\Phi : \mathcal{A} \to C(\widehat{\mathcal{A}}_1)$ be the Gelfand transform. Then for any $T \in \mathcal{A}$, image$(\Phi(T)) \supset \sigma(T)$.*

PROOF: By 4.3.6, we have $\widehat{\mathcal{A}} = \widehat{\mathcal{A}}_1$. If $\alpha \in \sigma(T)$, then $T - \alpha I$ is not invertible, and hence, by 4.3.7, there is $\lambda \in \widehat{\mathcal{A}}$ such that $\lambda(T - \alpha I) = 0$, i.e., $\lambda(T) = \alpha\lambda(I) = \alpha$. Thus $(\Phi(T))(\lambda) = \alpha$.

REMARK 4.3.10: In fact, one can easily show range $\Phi(T) = \sigma(T)$, although we do not use this. (See exercise 4.11.)

We can now prove Theorem 4.3.3.

PROOF OF 4.3.3: By Lemma 4.3.6, we have $\widehat{\mathcal{A}} = \widehat{\mathcal{A}}_1$ and $\Phi : \mathcal{A} \to C(\widehat{\mathcal{A}})$ is an algebra homomorphism with $\Phi(T^*) = \overline{\Phi(T)}$ and $\Phi(I) = 1$. By definition of $\widehat{\mathcal{A}}_1$, we have $\|\Phi(T)\| \leq \|T\|$. We claim in fact that this is an equality. By Corollary 4.3.9, we have $\|\Phi(T)\| \geq \|T\|_\sigma$, and by 4.1.15, since any $T \in \mathcal{A}$ is normal, we have $\|T\|_\sigma = \|T\|$. Thus, $\|\Phi(T)\| = \|T\|$.

To complete the proof, it therefore suffices to see that Φ is surjective, and since Φ is an isometry of Banach spaces, to see that $\Phi(\mathcal{A})$ is dense in $C(\widehat{\mathcal{A}})$. Since $\Phi(\mathcal{A})$ is a subalgebra of $C(\widehat{\mathcal{A}})$ containing 1, is closed under conjugation, and separates points by definition, density follows by Stone-Weierstrass (A.8).

COROLLARY 4.3.11. (Spectral theorem for commutative C^*-algebras) *Let E be a Hilbert space and $\mathcal{A} \subset B(E)$ a commutative C^*-algebra, $I \in \mathcal{A}$. Then the conclusions of Theorem 4.2.13 hold.*

COROLLARY 4.3.12. *(Spectral theorem for normal operators). If T is a normal operator, then T is unitarily equivalent to a multiplication operator.*

4.4. Mean ergodic theorem

We now apply the spectral theorem to obtain a general result about unitary operators. This result had its origins in studying the translation operator defined by a finite measure preserving homeomorphism. We first discuss this situation.

Let X be a compact metric space and $\varphi : X \in X$ a homeomorphism. Suppose $\mu \in M(X)$ is φ- invariant. We recall (Definition 2.3.7) that μ is called ergodic for φ if $A \subset X$ is measurable and $\varphi(A) = A$ implies $\mu(A) = 0$ or $\mu(A) = 1$. Since we shall now take μ as fixed, we shall then simply say that φ is ergodic.

DEFINITION 4.4.1. *φ is called mixing if for all measurable $A, B \subset X$ (say with $\mu(B) > 0$), we have*

$$\lim_{n \to \infty} \mu(\varphi^n(A) \cap B)/\mu(B) = \mu(A).$$

Heuristically, this means that for any B, the proportion of B taken up by $\varphi^n(A)$ is approximately $\mu(A)$, as long as n is large. Thus, by repeated application, φ spreads A throughout all of X in a very even manner.

PROPOSITION 4.4.2. *If φ is mixing, then φ is ergodic.*

PROOF: If $\mu(A) > 0$, then $\mu(\varphi^n A \cap A)/\mu(A) \to \mu(A)$. If $\varphi(A) = A$, we have $\mu(A) = 1$.

EXAMPLE 4.4.3: The converse of 4.4.2 is false. For example, if $\alpha/2\pi \notin Q$, then rotation by α is ergodic (Example 2.3.8 and exercise 2.13). However, it is easy to see that it is not mixing (Exercise 4.9).

We can understand a difference between mixing and ergodicity by viewing the former as a strong, quantitative form of the latter, which is fundamentally a qualitative notion. It is however a basic result that although ergodicity does not imply mixing, it does imply a weaker quantitative property.

DEFINITION 4.4.4. φ is called mixing in mean if for all measurable $A, B \subset X$ with $\mu(B) > 0$, $\mu(\varphi^n(A) \cap B)/\mu(B)$ is Cesaro convergent to $\mu(A)$. That is,

$$\lim_{N \to \infty} \frac{1}{N} \sum_{n=0}^{N-1} \mu(\varphi^n(A) \cap B)/\mu(B) = \mu(A).$$

THEOREM 4.4.5. (Mean ergodic theorem, I) φ is ergodic if and only if it is mixing in mean.

To prove 4.4.5, we first prove a general result about unitary operators.

THEOREM 4.4.6. (Mean ergodic theorem, II; von Neumann) Let E be a Hilbert space and $U \in B(E)$ a unitary operator. Let $E_0 = \{x \in E \mid Ux = x\}$ and $P_0 : E \to E_0$ be orthogonal projection. Then

$$\frac{1}{N} \sum_{n=0}^{N-1} U^n \to P_0$$

in the strong operator topology.

PROOF: By the spectral theorem we can assume $E = L^2(X)$ and $U = M_\psi$, where $\psi \in L^\infty(X)$ with $|\psi(x)| = 1$ a.e.. We then must have that $E_0 = \{f \in L^2(X) \mid f(x) = 0$ for (a.e.) x such that $\psi(x) \neq 1\}$. Therefore $P_0 = M_{\chi_{\psi^{-1}(1)}}$. Hence, it suffices to see that

$$M_{\frac{1}{N} \sum_{n=0}^{N-1} \psi^n} \longrightarrow M_{\chi_{\psi^{-1}(1)}}$$

in the strong operator topology. That is, for all $f \in L^2(X)$

$$\left\| \left(\frac{1}{N} \sum_{n=0}^{N-1} \psi^n \right) f - \chi_{\psi^{-1}(1)} f \right\|_2 \to 0.$$

Since $f \in L^2$ and

$$\left| \frac{1}{N} \sum_{n=0}^{N-1} \psi^n - \chi_{\psi^{-1}(1)} \right| \leq 2,$$

this will follow by the dominated convergence theorem if we can show

$$\frac{1}{N} \sum_{n=0}^{N-1} \psi^n \to \chi_{\psi^{-1}(1)} \text{ a.e.}$$

Since $|\psi(x)| = 1$, it suffices to see that if $|z| = 1$, then

$$\lim_{N \to \infty} \frac{1}{N} \sum_{n=0}^{N-1} z^n = \begin{cases} 0 & z \neq 1 \\ 1 & z = 1. \end{cases}$$

But if $z \neq 1$, then

$$\frac{1}{N} \sum_{n=0}^{N-1} z^n = \frac{1}{N} \frac{1 - z^N}{1 - z},$$

and since $|z^N| = 1$, this clearly converges to 0 as $N \to \infty$.

REMARK: Theorem 4.4.6 has a more elementary proof, i.e., one that does not involve an application of the spectral theorem. However, the argument we have given here shows clearly how the spectral theorem reduces a question about an operator to one about a function. In this case, the latter question is easily answered.

To deduce 4.4.5 from 4.4.6, we need the following simple remark.

LEMMA 4.4.7. *If $\varphi : X \to X$ is ergodic and $f : X \to \mathbf{C}$ is measurable with $f \circ \varphi = f$ a.e., then f is constant (a.e.).*

PROOF: By considering real and imaginary parts, we may assume $f(X) \subset \mathbf{R}$. Let

$$X_1 = \bigcap_{n \in \mathbf{Z}} \{x \mid f(\varphi^n(x)) = f(x)\}$$

Then $\mu(X_1) = 1, \varphi(X_1) = X_1$, and for $x \in X_1$ we have $f(\varphi(x)) = f(x)$. If f is not constant a.e., then there is some $\alpha \in \mathbf{R}$ such that

$0 < \mu(f^{-1}(-\infty, \alpha)) < 1$, and hence $0 < \mu(X_1 \cap f^{-1}(-\infty, \alpha)) < 1$. However, we clearly have $\varphi(X_1 \cap f^{-1}(-\infty, \alpha)) = X_1 \cap f^{-1}(-\infty, \alpha)$, contradicting ergodicity.

PROOF OF THEOREM 4.4.5: Letting $B = A$ (as in the proof of 4.4.2) we see that mixing in mean implies ergodicity. For the converse, let U be the unitary operator given by translation by φ. That is, $(Uf)(x) = f(\varphi^{-1}(x))$. We want to see that for all A, B,

$$\frac{1}{N} \sum_{n=0}^{N-1} \mu(\varphi^n(A) \cap B) \to \mu(A)\mu(B)$$

We convert this to an assertion about U. Since $\mu(A) = \langle \chi_A, 1 \rangle$, $\mu(B) = \langle 1, \chi_B \rangle$, and

$$\mu(\varphi^n(A) \cap B) = \langle \chi_{\varphi^n(A)}, \chi_B \rangle = \langle U^n \chi_A, \chi_B \rangle,$$

it suffices to see that for all $f, g \in L^2(X)$ we have

$$\langle \frac{1}{N} \sum_{n=0}^{N-1} U^n f, g \rangle \to \langle f, 1 \rangle \langle 1, g \rangle$$

By 4.4.6 and 4.4.7, we have

$$\frac{1}{N} \sum_{n=0}^{N-1} U^n f \to \langle f, 1 \rangle 1$$

where $\langle f, 1 \rangle 1 = P_0 f$ is projection onto the constant functions. Simply taking the inner product with g yields the desired result.

REMARK: One can formulate the theory of stationary stochastic processes in terms of a measure preserving transformation of a space with a probability measure, together with a function (i.e. random variable) on this space. The mean ergodic theorem applied to this situation then yields the weak law of large numbers.

Problems for Chapter 4

4.1 If E is a Hilbert space (over $\mathbb{C}$) and $T \in B(E)$ with $\langle Tx, x \rangle = 0$ for all $x \in E$, show $T = 0$.

4.2 If $T \in B(E)$, show T is normal if and only if $\|Tx\| = \|T^*x\|$ for all $x \in E$.

4.3 Show that every $T \in B(E)$ can be uniquely expressed as $T = T_1 + iT_2$ where $T_i \in B(E)$ with $T_i^* = T_i$. Show that T is normal if and only if $T_1 T_2 = T_2 T_1$.

4.4 If X is a measure space and $\varphi \in L^\infty(X)$, let M_φ be multiplication by φ on $L^2(X)$. If $\mu \sim \nu$ (i.e., they have the same null sets), show that there is a unitary operator $U : L^2(X, \mu) \to L^2(X, \nu)$ such that for any $\varphi \in L^\infty(X, \mu)$ $(= L^\infty(X, \nu))$ we have $U^{-1} M_\varphi U = M_\varphi$.

4.5 What is the spectrum of the operator in problem 3.10(a)?

4.6 Let $\{e_i\}$ be an orthonormal basis of E. Suppose $T \in B(E)$ is strictly upper triangular, i.e., $\langle Te_j, e_i \rangle = 0$ if $i \geq j$.

   a) If $\dim E < \infty$, show $\sigma(T) = \{0\}$.

   b) Give an example to show that we need not have $\sigma(T) = \{0\}$ if $\dim E = \infty$. Hint: If $S_n = T | \text{span}\{e_1, \ldots, e_n\}$, what is $(I - S_n)^{-1}$?

4.7 a) Give an example of two self-adjoint operators $T_1, T_2 \in B(E)$, with cyclic vectors, such that $\sigma(T_1) = \sigma(T_2)$, but T_1 is not unitary equivalent to T_2.

   b) If we further assume T_i are compact and injective, show T_1 is unitary equivalent to T_2.

4.8 If $T = T^*$, show that $T = \lim T_n$ (norm topology) where each T_n is of the form $\sum_{j=1}^{N} c_j P_j$, where $c_j \in \mathbb{C}$ and P_j is an orthogonal projection operator in E with $P_i(E) \perp P_j(E)$ for $i \neq j$, and $P_j T = T P_j$ for all j.

4.9 Let $\varphi : S^1 \to S^1$ be given by $\varphi(z) = e^{i\alpha} z$ where $\alpha \in \mathbf{R}$ is fixed. Show that φ is not mixing.

4.10 Prove Corollary 3.2.5 in general.

4.11 If $\mathcal{A} \subset B(E)$ is a commutative C^*-algebra with $I \in \mathcal{A}$, and Φ is the Gelfand transform, show that for any $T \in \mathcal{A}$ we have $\text{image}(\Phi(T)) = \sigma(T)$

4.12 If $T \in B(E)$, show that T is a linear combination of at most 4 unitary operators.

4.13 If $T \in B(E)$ is normal, T is called positive if $\langle Tx, x \rangle \geq 0$ for all $x \in E$. For T normal, show the following are equivalent:

i) T is positive.

ii) $\sigma(T) \subset [0, \infty)$.

iii) $T = S^*S$ for some $S \in B(E)$.

iv) $T = A^2$ for some self-adjoint $A \in B(E)$.

4.14 Let X be a compact Hausdorff space. For $x \in X$ let $e_x :$ $C(X) \to \mathbb{C}$ be $e_x(f) = f(x)$. Show that $\ker(e_x)$ is a maximal ideal in $C(X)$ and that every maximal ideal is of this form.

4.15 Suppose $\mathcal{A} \subset B(E)$ is a C^*-algebra, with $I \in \mathcal{A}$. For $T \in \mathcal{A}$, let $\sigma(T)$ denote the spectrum (as element of $B(E)$) and $\sigma_{\mathcal{A}}(T)$ denote the spectrum as an element of $\mathcal{A}$, so that $\sigma(T) \subset \sigma_{\mathcal{A}}(T)$.

i) Show boundary$(\sigma_{\mathcal{A}}(T)) = $ boundary$(\sigma(T))$.

ii) If $T = T^*$, show $\sigma_{\mathcal{A}}(T) = \sigma(T)$. (Hint: $\sigma(T) \subset \mathbb{R}$.)

iii) Show $\sigma_{\mathcal{A}}(T) = \sigma(T)$ for any $T \in \mathcal{A}$. (Hint: If $S \in \mathcal{A}$ and S is invertible in $B(H)$, so is S^*S.)

4.16 Let A be a Banach algebra with identity, and $B(A)$ the Banach algebra of bounded linear maps on A. If $x \in A$, let $m_x \in B(A)$ be $m_x(y) = xy$. Show that $\sigma_A(x) = \sigma_{B(A)}(m_x)$.

4.17 Prove the following version of the mean ergodic theorem. Let U be a unitary operator and P the orthogonal projection onto the space of U-invariant vectors. Show

$$\frac{1}{n-m} \sum_{j=m}^{n-1} U^j \to P$$

in the strong operator topology as $(n - m) \to \infty$.

FOURIER TRANSFORMS AND
SOBOLEV EMBEDDING THEOREMS

5.1 Basic properties of the Fourier transform and the Plancherel theorem

In this section we introduce the Fourier transform for functions on $\mathbf{R}^n$. This is an enormously useful and flexible tool. In particular, we shall apply it in section 5.2 to prove the Sobolev embedding theorem, and see in this section how it provides an explicit unitary equivalence of the unitary operators given by the regular representation of $\mathbf{R}^n$ with a family of multiplication operators. (The existence of such an equivalence is guaranteed by the spectral theorem (4.3.11).)

For ease of notation we let dx be Lebesgue measure on $\mathbf{R}^n$, but usually use the measure $dm = (2\pi)^{-n/2}\,dx$. The reason for this choice of normalization (which of course is not necessary but merely a convenience) will become apparent later. We remark that we may still use Fubini's theorem freely since $dm_{n_1} \times dm_{n_2} = dm_{n_1+n_2}$. Also for notational convenience we consider two copies of Euclidean space, which we denote by $\mathbf{R}^n$, $\widehat{\mathbf{R}}^n$. For $x \in \mathbf{R}^n$, $\xi \in \widehat{\mathbf{R}}^n$, $x = (x_1, \dots, x_n)$, $\xi = (\xi_1, \dots, \xi_n)$, we set $(x, \xi) = \sum x_i \xi_i$. (Thus, it may (or may not) be helpful to think of $\widehat{\mathbf{R}}^n$ as the dual of $\mathbf{R}^n$.) For functions f, g defined on $\mathbf{R}^n$, we let $\langle f, g \rangle = \int f\overline{g}\,dm$ whenever fg is integrable.

DEFINITION 5.1.1. *If $f \in L^1(\mathbf{R}^n)$, we define the Fourier transform of f, denoted by $\hat{f}$ or $\mathcal{F}(f)$, by*

$$\widehat{f}(\xi) = \int_{\mathbf{R}^n} f(x)e^{-i(x,\xi)}dm(x) = \langle f, e^{i(\cdot,\xi)}\rangle.$$

PROPOSITION 5.1.2. $\mathcal{F} : L^1(\mathbf{R}^n) \to BC(\widehat{\mathbf{R}}^n)$, and as a map between these Banach spaces $\mathcal{F}$ is bounded with $\|\mathcal{F}\| \le 1$.

PROOF: Clearly $|\hat{f}(\xi)| \le \int |f| = \|f\|_1$. Therefore $\hat{f}$ is bounded on $\widehat{\mathbf{R}}^n$, with bound $\|f\|_1$. Furthermore

$$\hat{f}(\xi + h) - \hat{f}(\xi) = \int f(x)e^{-i(x,\xi)}(e^{-i(x,h)} - 1)\, dm(x).$$

Therefore, as $h \to 0$ the dominated convergence theorem implies $\hat{f}(\xi + h) - \hat{f}(\xi) \to 0$, so that $\hat{f}$ is continuous.

In fact, for $f \in L^1(\mathbf{R}^n)$, $\hat{f}$ is not only bounded but vanishes at ∞. We define $C_0(\widehat{\mathbf{R}}^n) \subset BC(\widehat{\mathbf{R}}^n)$ by

$$C_0(\widehat{\mathbf{R}}^n) = \{h \in C(\widehat{\mathbf{R}}^n) \mid h(\xi) \to 0 \text{ as } \xi \to \infty\}.$$

PROPOSITION 5.1.3. (Riemann-Lebesgue lemma) If $f \in L^1(\mathbf{R}^n)$, then $\hat{f} \in C_0(\widehat{\mathbf{R}}^n)$. Thus, $\mathcal{F} : L^1(\mathbf{R}^n) \to C_0(\widehat{\mathbf{R}}^n)$ is bounded with $\|\mathcal{F}\| \le 1$.

PROOF: We let ρ denote the regular representation of $\mathbf{R}^n$ on any $L^p(\mathbf{R}^n)$. Thus $(\rho(t)f)(x) = f(x - t)$. In the equation

$$\hat{f}(\xi) = \int f(x)e^{-i(x,\xi)}dm(x),$$

make the change of variables $x \mapsto x - \frac{\pi\xi}{\|\xi\|^2}$. We obtain

$$\hat{f}(\xi) = \int \left[\rho\left(\frac{\pi\xi}{\|\xi\|^2}\right)f\right](x)e^{-i(x,\xi)}e^{\pi i}\, dm(x).$$

Adding this to the defining equation of $\hat{f}(\xi)$ we obtain

$$2\hat{f}(\xi) = \int \left(f - \rho\left(\frac{\pi\xi}{\|\xi\|^2}\right)f\right)(x)e^{-i(x,\xi)}\, dm(x),$$

and hence

$$2|\hat{f}(\xi)| \le \|f - \rho\left(\frac{\pi\xi}{\|\xi\|^2}\right)f\|_1.$$

As $\xi \to \infty$, $\frac{\pi\xi}{\|\xi\|^2} \to 0$, so the result follows by 1.3.10.

One somewhat unsatisfactory feature of the conclusion of 5.1.3 is that a function in $C_0(\widehat{\mathbf{R}}^n)$ need not be in any $L^p(\widehat{\mathbf{R}}^n)$ for $p < \infty$. Although we shall return later to the question of the integrability properties of $\hat{f}$ in more detail, here we wish only to see that for a large class of L^1-functions f, we will have $\hat{f}$ in L^p. We begin with a general remark on integrability. For $n = 1$, one easily computes that

$$\int_{\widehat{\mathbf{R}}} \frac{1}{(1 + \xi^2)} \in L^1(\widehat{R}).$$

Hence, for any n, we have

$$\frac{1}{\prod_1^n (1 + \xi_i^2)} \in L^1(\widehat{\mathbf{R}}^n)$$

Therefore, if h is a measurable function on $\widehat{\mathbf{R}}^n$ with $\left(\prod_1^n (1 + \xi_i^2)\right)h(\xi)$ bounded, it follows that $h \in L^1(\widehat{\mathbf{R}}^n)$.

PROPOSITION 5.1.4.

a)   If $f \in C_c^1(\mathbf{R}^n)$ (so that $\frac{\partial f}{\partial x_j} \in C_c(\mathbf{R}^n)$), then

$$\widehat{\frac{\partial f}{\partial x_j}}(\xi) = -i\xi_j\,\hat{f}(\xi)$$

b)   If $f \in C_c^k(\mathbf{R}^n)$, then for any $|\alpha| \le k$

$$(\widehat{D^\alpha f})(\xi) = (-i)^{|\alpha|}\xi^\alpha\,\hat{f}(\xi).$$

c)   If $f \in C_c^\infty(\mathbf{R}^n)$, then $\hat{f} \in C_0(\widehat{\mathbf{R}}^n) \cap L^p(\widehat{\mathbf{R}}^n)$ for every $p \ge 1$.

PROOF: (a) follows via integration by parts. Namely,

$$\frac{\widehat{\partial f}}{\partial x_j}(\xi) = \int \frac{\partial f}{\partial x_j}(x)e^{-i(x,\xi)}\,dm(x)$$

$$= -\int f(x)\frac{\partial(e^{-i(x,\xi)})}{\partial x_j}\,dm(x)$$

$$= -i\xi_j \int f(x)e^{-i(x,\xi)}\,dm(x)$$

$$= -i\xi_j \hat{f}(\xi).$$

(b) follows from (a) by induction. To verify (c), we observe that since $D^\alpha f \in C_c^\infty(\mathbf{R}^n) \subset L^1(\mathbf{R}^n)$, Proposition 5.1.2 and (b) imply that $\xi^\alpha \hat{f}(\xi)$ is bounded for all α, and hence so is $\prod_j(1 + \xi_j^2)\hat{f}(\xi)$. It follows from our remarks above that $\hat{f} \in L^1(\widehat{\mathbf{R}}^n)$. Since we also have $\hat{f} \in C_0(\widehat{\mathbf{R}}^n)$ by 5.1.3, it follows that $\hat{f} \in L^p(\widehat{\mathbf{R}}^n)$ for all $p \geq 1$.

We observe that (c) shows that for at least a dense subset in $L^1(\mathbf{R}^n)$ we have $\hat{f} \in L^1(\widehat{\mathbf{R}}^n)$. Conclusion (b), however, is of interest far beyond using it to verify (c). Namely, it shows that under $\mathcal{F}$, the operator D^α is converted into a simple multiplication. This will clearly be more useful if we can recapture f from $\hat{f}$, i.e., invert $\mathcal{F}$. It is to this issue that we now turn. First, we collect some simple facts.

LEMMA 5.1.5. Let ρ be the regular representation of $\mathbf{R}^n$ (or $\widehat{\mathbf{R}}^n$); i.e., $(\rho(t)f)(x) = f(x - t)$ (and similarly for $\widehat{\mathbf{R}}^n$). For each $a \in \mathbf{R}$, $a \neq 0$, define $(S_a f)(x) = f(x/a)$ where f is a function on $\mathbf{R}^n$, and similarly define $(S_a h)(\xi) = h(\xi/a)$ where h is a function on $\widehat{\mathbf{R}}^n$. Then:

    i) $(\widehat{\rho(t)f})(\xi) = e^{-i(t,\xi)}\hat{f}(\xi).$
    ii) $(\rho(t)\hat{f})(\xi) = (e^{i(\cdot,t)}f)\hat{}(\xi).$
    iii) $\widehat{S_a f}(\xi) = a^n(S_{1/a}\hat{f})(\xi).$
    iv) $(S_a \hat{f})(\xi) = a^n(S_{1/a}f)^\wedge(\xi).$

PROOF: Straightforward, using at most a change of variables.

We will also need the following sample calculation of a Fourier transform.

LEMMA 5.1.6. Let $\varphi(x) = e^{-\|x\|^2/2}$. Then $\widehat{\varphi} = \varphi$.

PROOF: We claim

$$e^{-\|\xi\|^2/2} = \int_{\mathbb{R}^n} e^{-\|x\|^2/2} e^{-i(x,\xi)} \, dm(x).$$

Writing this as a product of such equalities, it suffices to verify this for $n = 1$. Then

$$\begin{aligned}
\int_{\mathbb{R}} e^{-x^2/2} e^{-ix\xi} \frac{dx}{\sqrt{2\pi}} &= \int e^{-\frac{1}{2}x(x+2i\xi)} \frac{dx}{\sqrt{2\pi}} \\
&= \int_{\mathbb{R}+i\xi} e^{-\frac{1}{2}(z-i\xi)(z+i\xi)} \frac{dz}{\sqrt{2\pi}} \\
&= \int_{\mathbb{R}+i\xi} e^{-\frac{1}{2}(z^2+\xi^2)} \frac{dz}{\sqrt{2\pi}} \\
&= e^{-\xi^2/2} \int_{\mathbb{R}+i\xi} e^{-z^2/2} \frac{dz}{\sqrt{2\pi}} \\
&= e^{-\xi^2/2} \lim_{n\to\infty} \int_{-N+i\xi}^{N+i\xi} e^{-z^2/2} \frac{dz}{\sqrt{2\pi}}
\end{aligned}$$

Since $e^{-z^2/2}$ is analytic, the integral along a closed curve is 0, and hence the above expression

$$= e^{-\xi^2/2} \lim_{N\to\infty} \left(\int_{-N}^{N} e^{-z^2/2} \frac{dz}{\sqrt{2\pi}} + \int_{C_1\cup C_2} e^{-z^2/2} \frac{dz}{\sqrt{2\pi}} \right)$$

where C_1 is the straight line from $-N$ to $-N+i\xi$ and C_2 is the line from $N + i\xi$ to N. Since $|e^{-z^2/2}| \to 0$ uniformly as $|\mathrm{Re}(z)| \to \infty$, we obtain

$$e^{-\xi^2/2} \int_{-\infty}^{\infty} e^{-z^2/2} \frac{dz}{\sqrt{2\pi}}.$$

It is a well known calculus exercise that this latter integral is 1.

REMARK: It is exactly at this point that the particular choice of normalization of Lebesgue measure becomes relevant. Otherwise, we would have $\widehat{\varphi} = c\varphi$ for some constant c.

We now define the inverse Fourier transform.

DEFINITION 5.1.7. If $h \in L^1(\widehat{\mathbf{R}}^n)$, define the inverse Fourier transform $\tilde{h}$ of h by

$$\tilde{h}(x) = \int_{\widehat{\mathbf{R}}^n} h(\xi)e^{i(x,\xi)}dm(\xi).$$

We also denote $\tilde{h}$ by $\mathcal{F}^*(h)$.

Clearly, all the results we have stated so far for $\mathcal{F}$ have analogues for $\mathcal{F}^*$. One can either repeat essentially the same computation for $\mathcal{F}^*$, of more simply observe that $\tilde{h} = \overline{\overline{\hat{h}}}$, so that in fact $\mathcal{F}^*$ is very simply described in terms of $\mathcal{F}$ itself.

The first property we observe about $\mathcal{F}^*$ is that it behaves like an adjoint.

LEMMA 5.1.8. If $f \in L^1(\mathbf{R}^n)$ and $h \in L^1(\widehat{\mathbf{R}}^n)$, then $\langle \mathcal{F}(f), h \rangle = \langle f, \mathcal{F}^*(h) \rangle$.

PROOF: By Fubini's theorem we have

$$\begin{aligned}
\langle \mathcal{F}(f), h \rangle &= \int_{\widehat{\mathbf{R}}^n} \hat{f}(\xi)\bar{h}(\xi) \\
&= \int_{\widehat{\mathbf{R}}^n} \int_{\mathbf{R}^n} f(x)e^{-i(x,\xi)}\bar{h}(\xi) \\
&= \int_{\mathbf{R}^n} f(x)\left(\overline{\int_{\widehat{\mathbf{R}}^n} h(\xi)e^{i(x,\xi)}}\right) \\
&= \langle f, \mathcal{F}^*(h) \rangle.
\end{aligned}$$

We now use 5.1.8 to establish:

THEOREM 5.1.9. If $f \in C_c^\infty(\mathbf{R}^n)$, then $\mathcal{F}^*\mathcal{F}f = f$.

PROOF: We first observe that it suffices to show $(\mathcal{F}^*\mathcal{F}f)(0) = f(0$ for all $f \in C_c^\infty(\mathbf{R}^n)$. To see this, for any $t \in \mathbf{R}^n$ we apply Lemma 5.1.5. Thus, assuming equality at $t = 0$, we have for any t that

$$\begin{aligned}
f(t) &= \big(\rho(-t)f\big)(0) \\
&= \mathcal{F}^*\mathcal{F}\big(\rho(-t)f\big)(0) &&\text{(since } \rho(-t)f \in C_c^\infty(\mathbf{R}^n)\text{)} \\
&= \mathcal{F}^*\big(e^{i(t,\xi)}\mathcal{F}f\big)(0) &&\text{(by 5.1.5(i))} \\
&= \mathcal{F}^*\big(\mathcal{F}f\big)(t) &&\text{(by definition).}
\end{aligned}$$

To prove equality at 0, we observe that

$$(\mathcal{F}^*\mathcal{F}f)(0) = \int_{\widehat{\mathbf{R}}^n} \hat{f}(\xi)\, dm(\xi),$$

and hence we want to show $\langle \hat{f}, 1 \rangle = f(0)$ for any $f \in C_c^\infty(\mathbf{R}^n)$. Fix any function $\varphi \in C^\infty(\widehat{\mathbf{R}}^n) \cap L^1(\widehat{\mathbf{R}}^n)$ such that $1 \geq \varphi \geq 0$, and $\varphi(0) = 1$. Then as $a \to \infty$, $S_a \varphi \to 1$ pointwise (where $S_a \varphi(\xi) = \varphi(\xi/a)$ is as in 5.1.5). Furthermore, we clearly have $0 \leq S_a \varphi \leq 1$. Thus, by the dominated convergence theorem $\langle \hat{f}, S_a \varphi \rangle \to \langle \hat{f}, 1 \rangle$ as $a \to \infty$. By Lemma 5.1.8, we have $\langle \hat{f}, S_a \varphi \rangle = \langle f, \widetilde{S_a \varphi} \rangle$, and by 5.1.5(iii) for $\mathcal{F}^*$ we have

$$\langle \hat{f}, S_a \varphi \rangle = \langle f, a^n S_{1/a} \tilde{\varphi} \rangle = \int f(x) a^n \overline{\tilde{\varphi}}(ax)\, dm(x).$$

Making the change of variables $x \to x/a$, we obtain

$$\langle \hat{f}, S_a \varphi \rangle = \int f(x/a) \overline{\tilde{\varphi}}(x)\, dm(x)$$

Letting $a \to \infty$, the dominated convergence theorem implies that the integral converges to

$$f(0) \int \overline{\tilde{\varphi}}(x)\, dm(x).$$

Therefore, we deduce that for any $f \in C_c^\infty(\mathbf{R}^n)$,

$$\langle \hat{f}, 1 \rangle = f(0) \int \overline{\tilde{\varphi}}(x)\, dm(x).$$

(This already shows that $\mathcal{F}^*\mathcal{F}f = cf$ for some constant $c = \int \overline{\tilde{\varphi}}$, independent of f.)

To evaluate the constant $\int \overline{\tilde{\varphi}}\, dm$, we may choose any φ we like (satisfying the above conditions.). Taking $\varphi(\xi) = e^{-\|\xi\|^2/2}$, we have by Lemma 5.1.6 (and the fact that $\widetilde{\varphi} = \overline{\overline{\varphi}}$) that $\int \overline{\widetilde{\varphi}}\, dm = 1$. This completes the proof.

REMARK: With a different choice of normalization for Lebesgue measure, we would have $\mathcal{F}^*\mathcal{F}f = cf$ where $c \neq 1$. Cf. the remark following the proof of 5.1.6. We also clearly have $\mathcal{F}\mathcal{F}^*h = h$ for $h \in C_c^\infty(\widehat{\mathbf{R}}^n)$.

COROLLARY 5.1.10. For $f,g \in C_c^\infty(\mathbf{R}^n)$, we have $\langle \mathcal{F}f, \mathcal{F}g \rangle = \langle f,g \rangle$.

PROOF: $g \in C_c^\infty(\mathbf{R}^n)$ implies $\mathcal{F}g \in L^1(\widehat{\mathbf{R}}^n)$, so

$$\begin{aligned}\langle \mathcal{F}f, \mathcal{F}g \rangle &= \langle f, \mathcal{F}^*\mathcal{F}g \rangle && \text{by 5.1.8}\\ &= \langle f,g \rangle && \text{by 5.1.9.}\end{aligned}$$

THEOREM 5.1.11. (Plancherel) The map $\mathcal{F} : L^1(\mathbf{R}^n) \cap L^2(\mathbf{R}^n) \to C_0(\widehat{\mathbf{R}}^n)$ extends uniquely to a unitary operator $\mathcal{F} : L^2(\mathbf{R}^n) \to L^2(\widehat{\mathbf{R}}^n)$.

PROOF: By Corollary 5.1.10, $\|\mathcal{F}f\|_2 = \|f\|_2$ for $f \in C_c^\infty(\mathbf{R}^n)$. Thus, $\mathcal{F}\,|C_c^\infty(\mathbf{R}^n)$ extends uniquely to an isometry $\widetilde{\mathcal{F}} : L^2(\mathbf{R}^n) \to L^2(\widehat{\mathbf{R}}^n)$. We claim that if $f \in L^2(\mathbf{R}^n) \cap L^1(\mathbf{R}^n)$, then $\widetilde{\mathcal{F}}(f) = \hat{f}$. To see this, choose $f_n \in C_c^\infty(\mathbf{R}^n)$ such that $f_n \to f$ in both $L^1(\mathbf{R}^n)$ and $L^2(\mathbf{R}^n)$. (0.B.5) Then $\widetilde{\mathcal{F}}f_n \to \widetilde{\mathcal{F}}f$ in L^2 and $\hat{f}_n \to \hat{f}$ uniformly by Proposition 5.1.2. Since $\widetilde{\mathcal{F}}f_n = \hat{f}_n$, it follows that $\hat{f} = \widetilde{\mathcal{F}}f$.

To complete the proof, it suffices to see that $\widetilde{\mathcal{F}}$ is surjective, and since it is an isometry, it suffices to see the image is dense (Lemma 4.1.5). But if $h \in C_c^\infty(\widehat{\mathbf{R}}^n)$, $\widetilde{h} \in L^1(\mathbf{R}^n) \cap L^2(\mathbf{R}^n)$ (by 5.1.4) and $\mathcal{F}\widetilde{h} = h$ by Theorem 5.1.9. Thus, the range of $\widetilde{\mathcal{F}}$ contains $C_c^\infty(\widehat{\mathbf{R}}^n)$ which is dense in $L^2(\widehat{\mathbf{R}}^n)$.

The map $\mathcal{F} : L^2(\mathbf{R}^n) \to L^2(\widehat{\mathbf{R}}^n)$ is called the L^2-Fourier transform.

As above, let $\rho(t) \in U(L^2(\mathbf{R}^n))$ denote the regular representation of $\mathbf{R}^n$. Then $\{\rho(t) \mid t \in \mathbf{R}^n\}$ is a commuting family of unitary operators. By the spectral theorem (4.3.11) the operators are simultaneously unitary equivalent to multiplication operators on some L^2-space. In fact, the Plancherel theorem gives us an explicit realization of this.

PROPOSITION 5.1.12. *Let $\mathcal{F}$ be the L^2-Fourier transform. Then for any $t \in \mathbf{R}^n$, $\mathcal{F}\rho(t)\mathcal{F}^{-1} = M_{e^{-i(t,\cdot)}}$, where the latter is multiplication by the L^∞-function $e^{-i(t,\xi)}$ on $L^2(\widehat{\mathbf{R}}^n)$.*

PROOF: We have $\mathcal{F}(\rho(t)f) = M_{e^{-i(t,\cdot)}}(\mathcal{F}f)$ for all $f \in L^1 \cap L^2$ by 5.1.5. By continuity, this holds for all $f \in L^2(\mathbf{R}^n)$.

5.2 Sobolev Embedding Theorems

In this section we use the Fourier transform to show how the existence of sufficiently many weak derivatives in L^2 implies the existence of some derivatives in the classical sense.

We recall the notation ∂_j for j-th partial derivative, ∂_j^w for weak j-th partial derivative, D^α for α-th derivative and D_w^α for weak α-th derivative. We shall also use the notation $\hat{f}$ for $\mathcal{F}(f)$ where $f \in L^2$.

In Proposition 5.1.4 we saw the basic fact that if $f \in C_c^k(\mathbf{R}^n)$ and $|\alpha| \leq k$, then $(\widehat{D^\alpha f})(\xi) = (-i)^{|\alpha|}\xi^\alpha \hat{f}(\xi)$. We now establish this for weak derivatives.

PROPOSITION 5.2.1. *If $f \in W^{2,k}(\mathbf{R}^n)$, then for $|\alpha| \leq k$ we have*

$$(\widehat{D_w^\alpha f})(\xi) = (-i)^{|\alpha|}\xi^\alpha \hat{f}(\xi).$$

In particular, $\xi^\alpha \hat{f} \in L^2(\widehat{\mathbf{R}}^n)$ for all $|\alpha| \leq k$.

PROOF: By induction (and using 1.1.22(a)), it suffices to see this for $|\alpha| = 1$, i.e., $D^\alpha = \partial_j$. Let $\varphi \in C_c^\infty(\mathbf{R}^n)$. Then

$$\begin{aligned}
\langle(\widehat{\partial_j^w f}), \widehat{\varphi}\rangle &= \langle \partial_j^w f, \varphi\rangle && \text{(by Plancherel)}\\
&= -\langle f, \partial_j \varphi\rangle\\
&= -\langle \hat{f}, (\widehat{\partial_j \varphi})\rangle && \text{(by Plancherel)}\\
&= \langle \hat{f}, i\xi_j \widehat{\varphi}\rangle\\
&= \langle -i\xi_j \hat{f}, \widehat{\varphi}\rangle.
\end{aligned}$$

Since $\{\widehat{\varphi} \mid \varphi \in C_c^\infty(\mathbf{R}^n)\}$ is dense in $L^2(\widehat{\mathbf{R}}^n)$, the result follows.

Proposition 5.2.1 shows that for $f \in L^2(\mathbf{R}^n)$, the existence of weak derivatives implies integrability properties of $\hat{f}$.

We now consider the differentiability properties of $\hat{f}$.

PROPOSITION 5.2.2. *Suppose* $f \in L^1(\mathbf{R}^n)$ *and for all* $|\alpha| \le k$, $x^\alpha f \in L^1(\mathbf{R}^n)$. *Then* $\hat{f} \in BC^k(\mathbf{R}^n)$ *and*

$$(D^\alpha \hat{f})(\xi) = (-i)^{|\alpha|}(\widehat{x^\alpha f})(\xi).$$

PROOF: By an induction argument, it suffices to see this for $|\alpha| = 1$, i.e., $x^\alpha = x_j$. We have

$$\hat{f}(\xi) = \int_{\mathbf{R}^n} f(x)e^{-i(x,\xi)}\, dm(x).$$

Differentiating the integrand with respect to ξ_j, we obtain the expression

$$\int_{\mathbf{R}^n} f(x)(-ix_j)e^{-i(x,\xi)}\, dm(x).$$

By Lemma B.1, if this integral exists and is continuous in ξ, the integral must be $\frac{\partial \hat{f}}{\partial \xi_j}$. However, the integral is simply $(-i)(\widehat{x_j f})(\xi)$ which is continuous since $x_j f \in L^1(\mathbf{R}^n)$.

If we apply 5.2.2 to the inverse Fourier transform, we obtain:

COROLLARY 5.2.3. *Suppose* $f \in L^2(\mathbf{R}^n)$. *If* $\xi^\alpha \hat{f} \in L^1(\widehat{\mathbf{R}}^n)$ *for all* $|\alpha| \le k$, *then* $f \in BC^k(\mathbf{R}^n)$, *and* $D^\alpha f = i^{|\alpha|}(\xi^\alpha \hat{f})^\sim$.

PROOF: By 5.2.2 we have that $h \in L^1(\widehat{\mathbf{R}}^n)$ and $\xi^\alpha h \in L^1(\widehat{\mathbf{R}}^n)$ for $|\alpha| \le k$ implies $\tilde{h} \in BC^k(\mathbf{R}^n)$ and $(D^\alpha \tilde{h})(x) = i^{|\alpha|}(\widetilde{\xi^\alpha h})(x)$. Now let $h = \hat{f}$.

We can now prove the main theorem of this section

THEOREM 5.2.4. (Sobolev embedding Theorem) *If* $f \in W^{2,k}(\mathbf{R}^n)$ *and* $k > r + \frac{n}{2}$, *then* $f \in BC^r(\mathbf{R}^n)$. *Furthermore, the inclusion map* $W^{2,k}(\mathbf{R}^n) \to BC^r(\mathbf{R}^n)$ *is bounded.*

PROOF: (I): We first illustrate the argument by giving the proof for $k \ge r + n$. This avoids one technicality making the main point of the argument clearer. We then return to the case $k > r + \frac{n}{2}$.

Since $f \in W^{2,k}(\mathbf{R}^n)$, for all $|\alpha| \leq k$ we have $\xi^\alpha \hat{f} \in L^2(\widehat{\mathbf{R}}^n)$ (by 5.2.1). To see that $f \in BC^r(\mathbf{R}^n)$, it suffices by Corollary 5.2.3 to see that $\xi^\alpha \hat{f} \in L^1(\widehat{\mathbf{R}}^n)$ for $|\alpha| \leq r$. We do this by showing that $|\alpha| \leq r$ implies that $\xi^\alpha \hat{f}$ is a product of two functions in $L^2(\widehat{\mathbf{R}}^n)$. Namely, for any α we write $\xi^\alpha \hat{f} = h_1 h_2$ where

$$h_1 = \left[\prod (1 + |\xi_i|^{1+\alpha_i}) \right] \hat{f}$$

and

$$h_2 = \frac{\xi^\alpha}{\prod(1 + |\xi_i|^{1+\alpha_i})}$$

We claim $h_1, h_2 \in L^2(\widehat{\mathbf{R}}^n)$ as long as $|\alpha| \leq r$, where $k \geq r + n$. To see $h_1 \in L^2(\widehat{\mathbf{R}}^n)$, we observe that $|h_1|$ is dominated by a finite sum of terms of the form $|\xi_i|^\beta |\hat{f}|$ where

$$|\beta| \leq n + \sum \alpha_i \leq n + r \leq k.$$

Hence $h_1 \in L^2(\widehat{\mathbf{R}}^n)$. To see $h_2 \in L^2(\widehat{\mathbf{R}}^n)$ is simply an integration exercise. Namely, by Fubini's theorem it suffices to see that $\frac{x^n}{1+|x|^{1+n}} \in L^2(\mathbf{R})$ for any $n \geq 0$. Since this is continuous, it suffices to see it is in $L^2(\{|x| \geq 1\})$. However,

$$\frac{|x|^n}{1 + |x|^{1+n}} \leq \frac{1}{|x|},$$

which is square integrable on any closed set not containing 0.

To see that the embedding $W^{2,k}(\mathbf{R}^n) \to BC^r(\mathbf{R}^n)$ is bounded, we have, via the expression $\xi^\alpha \hat{f} = h_1 h_2$ that

$$\|\xi^\alpha \hat{f}\|_1 \leq \|h_1\|_2 \|h_2\|_2.$$

Since $\|h_2\|_2$ is independent of f, we have

$$\|\xi^\alpha \hat{f}\|_1 \leq C \cdot \sum_{|\beta| \leq k} \|\xi^\beta \hat{f}\|_2$$

for some constant C, where $|\alpha| \leq r$. By Corollary 5.2.3 $(\xi^\alpha \hat{f})^\sim = i^{|\alpha|} D^\alpha f$, and hence by 5.1.2 (applied to the inverse Fourier transform) we have

$$\begin{aligned}
\|D^\alpha f\|_\infty &\leq \|\xi^\alpha \hat{f}\|_1 \\
&\leq C \sum_{|\beta| \leq k} \|\xi^\beta \hat{f}\|_2 \\
&= C \sum_{|\beta| \leq k} \|D_w^\beta f\|_2
\end{aligned}$$

(by 5.2.1 and the Plancherel theorem)

$$= C\|f\|_{2,k}$$

II): The only change in the argument above needed to improve the result to $k > r + \frac{n}{2}$ is to change the decomposition $\xi^\alpha \hat{f} = h_1 h_2$ so that we have $h_i \in L^2(\widehat{\mathbf{R}}^n)$ for $|\alpha| < r$ where $k > r + \frac{n}{2}$, and h_2 is independent of f. We let

$$\begin{aligned}
h_1 &= (1 + \|\xi\|^k)\hat{f} \qquad \text{and} \\
h_2 &= \frac{\xi^\alpha}{1 + \|\xi\|^k}.
\end{aligned}$$

To see $h_1 \in L^2$, it suffices to see that $\|\xi\|^k \hat{f} \in L^2$, i.e., $\|\xi\|^{2k}|\hat{f}|^2 \in L^1(\widehat{\mathbf{R}}^n)$. However $\|\xi\|^{2k}|\hat{f}|^2$ is a sum of terms of the form $\xi^{2\beta}|\hat{f}|^2$ where $|\beta| \leq k$. Since $\xi^\beta \hat{f} \in L^2$, we have $\xi^{2\beta}|\hat{f}|^2 \in L^1$, showing that $h_1 \in L^2$. Furthermore, this shows that

$$\|h_1\|_2^2 \leq \sum_{|\beta| \leq k} \|\xi^\beta \hat{f}\|_2^2,$$

and hence (as in the proof of (I)), that

$$\|h_1\|_2^2 \le \sum_{|\beta| \le k} \|D_w^\beta f\|_2^2.$$

Thus, as in (I), to complete the proof we need only show that $h_2 \in L^2(\widehat{\mathbf{R}}^n)$. We have $|\xi^\alpha| \le \|\xi\|^r$ for $|\alpha| \le r$, so we need only show that $k > r + \frac{n}{2}$ implies

$$\frac{\|\xi\|^r}{1 + \|\xi\|^k} \in L^2(\widehat{\mathbf{R}}^n)$$

We can write $d\xi$ in "spherical coordinates" as $d\xi = cR^{n-1}dRdS$ where $R = \|\xi\|$ and $dS =$ standard measure on S^{n-1}. Thus it suffices to show

$$\int \frac{R^{2r}}{(1 + R^k)^2} R^{n-1} dR < \infty.$$

It suffices to show

$$\int_{|R| \ge 1} \frac{R^{2r} R^{n-1}}{R^{2k}} dR < \infty.$$

This is the case if $2r + n - 1 - 2k < -1$, i.e., $r + \frac{n}{2} < k$.

We now wish to extend the inclusion $W^{2,k}(\mathbf{R}^n) \hookrightarrow BC^r(\mathbf{R}^n)$ to general open sets $U \subset \mathbf{R}^n$.

COROLLARY 5.2.5. *Let $\Omega \subset \mathbf{R}^n$ be open and $f \in W^{2,k}(\Omega)$. If $k > r + \frac{n}{2}$, then $f \in C^r(\Omega)$.*

PROOF: Let $x \in \Omega$ and choose $\Omega_0 \subset \Omega$ open with $x \in \Omega_0 \subset \overline{\Omega}_0 \subset \Omega$, and $\overline{\Omega}_0$ compact. Let $\varphi \in C_c^\infty(\Omega)$ such that $\varphi = 1$ on Ω_0. Then $\varphi f \in W^{2,k}(\mathbf{R}^n)$ by Proposition 1.1.21, so we can apply 5.2.4 to deduce $\varphi f \in C^r(\Omega)$. This implies $f |\Omega_0$ is in $C^r(\Omega_0)$, and since $x \in \Omega$ is arbitrary, $f \in C^r(\Omega)$.

The question of continuity properties of the inclusion $W^{2,k}(\Omega) \hookrightarrow C^r(\Omega)$ is more delicate for general Ω than it is for $\mathbf{R}^n$. Satisfactory general theorems depend upon the nature of the boundary $\partial\Omega$, and we shall not discuss this here. However, we can easily deduce one such result for arbitrary Ω. Recall that $C_c^\infty(\Omega) \subset W^{2,k}(\Omega)$. We let $H^k(\Omega)$ denote the closure of $C_c^\infty(\Omega)$.

COROLLARY 5.2.6. *If $k > r + \frac{n}{2}$, we have a bounded inclusion $H^k(\Omega) \hookrightarrow BC^r(\Omega)$.*

PROOF: The inclusion $C_c^\infty(\Omega) \hookrightarrow C_c^\infty(\mathbf{R}^n)$ extends to a continuous embedding $H^k(\Omega) \to W^{2,k}(\mathbf{R}^n)$, and hence by Theorem 5.2.4 gives a bounded inclusion $i : H^k(\Omega) \to BC^r(\mathbf{R}^n)$. The inclusion $H^k(\Omega) \to BC^r(\Omega)$ is simply the composition of i with the restriction map $BC^r(\mathbf{R}^n) \to BC^r(\Omega)$.

REMARK 5.2.7: While we have stated and proved the Sobolev embedding theorem for $W^{2,k}$, there are also results for $W^{p,k}$. Namely, one can show that there is an inclusion $W^{p,k}(\mathbf{R}^n) \to C^r(\mathbf{R}^n)$ for $k > r + \frac{n}{p}$ (and $1 \le p < \infty$).

We now discuss conditions under which one has compactness of embeddings.

THEOREM 5.2.8. (Rellich) *Let $\Omega \subset \mathbf{R}^n$ be an open bounded set.*
   a) For any $k \ge 1$, the inclusion $H^k(\Omega) \hookrightarrow H^{k-1}(\Omega)$ is a compact operator.
   b) If $k > r + \frac{n}{2} + 1$, then the inclusion $H^k(\Omega) \hookrightarrow BC^r(\Omega)$ is a compact operator.

PROOF: (b) follows from (a) and Corollary 5.2.6. Therefore, we need only prove (a). Since $C_c^\infty(\Omega)$ is dense in $H^k(\Omega)$, it suffices to see that if $f_j \in C_c^\infty(\Omega)$ is a sequence which is bounded in $H^k(\Omega)$, then f_j has a Cauchy subsequence in $H^{k-1}(\Omega)$. In fact, it suffices to prove this for $k = 1$. Namely, if f_j is bounded in $H^k(\Omega)$ then $D^\alpha f_j$ is bounded in $H^1(\Omega)$ for all $|\alpha| \le k - 1$, and if the result is established for $k = 1$ we would have that $D^\alpha f_j$ has a Cauchy subsequence in $L^2(\Omega)$ for each such α. This easily implies that f_j has a Cauchy subsequence in $H^{k-1}(\Omega)$. We thus assume $k = 1$.

By the Plancherel theorem we have

$$\|f_j - f_k\|_2^2 = \|\hat{f}_j - \hat{f}_k\|_2^2 = \int_{\|\xi\| \le R} |\hat{f}_j - \hat{f}_k|^2 + \int_{\|\xi\| \ge R} |\hat{f}_j - \hat{f}_k|^2$$

for any $R > 0$. Since $\{f_j\}$ is bounded in $H^1(\Omega) \subset W^{2,1}(\mathbf{R}^n)$, we have for each i that $\{\partial_i f_j\}$ is bounded in $L^2(\mathbf{R}^n)$, and hence by the

Plancherel theorem that $\xi_i \hat{f}_j$ is bounded in $L^2(\hat{\mathbf{R}}^n)$. Hence $\{\|\xi\|\hat{f}_j\}$ is bounded in $L^2(\hat{\mathbf{R}}^n)$, say $\left\|\left(\|\xi\|\hat{f}_j\right)\right\|_2 \leq C$ for all j. Then

$$\int_{\|\xi\| \geq R} |\hat{f}_j - \hat{f}_k|^2 \leq \int_{\|\xi\| \geq R} \frac{\|\xi\|^2}{R^2} |\hat{f}_j - \hat{f}_k|^2 \, d\xi \leq \frac{C^2}{R^2}.$$

Thus, given any $\varepsilon > 0$, we may choose R so that

$$\int_{\|\xi\| \geq R} |\hat{f}_j - \hat{f}_k|^2 < \varepsilon/2$$

for all j, k.

We now turn to the integral over $\{\|\xi\| \leq R\}$. Since bounded sets in $L^2(\Omega)$ are weak-* compact (1.1.31), by passing to a subsequence we may assume f_j is weak-* convergent. Since Ω is bounded, for any ξ we have $e^{i(\cdot,\xi)} \in L^2(\Omega)$ and hence $\hat{f}_j(\xi) = \langle f_j, e^{i(\cdot,\xi)} \rangle$ is convergent, and in particular Cauchy. That is, $\hat{f}_j(\xi) - \hat{f}_k(\xi) \to 0$ pointwise as $j, k \to \infty$. Once again using the fact that Ω is bounded, we have a continuous inclusion $L^2(\Omega) \to L^1(\Omega)$. Thus $\{f_j\}$ is bounded in $L^1(\Omega)$, and hence $\{\hat{f}_j\}$ is uniformly bounded in $BC(\hat{\mathbf{R}}^n)$ (5.1.2). Thus, by the dominated convergence theorem

$$\int_{\|\xi\| \leq R} |\hat{f}_j(\xi) - \hat{f}_k(\xi)|^2 \to 0 \qquad \text{as } j, k \to \infty.$$

Therefore, for j, k sufficiently large, we have $\|f_j - f_k\|_2^2 < \varepsilon$, proving the theorem.

The Sobolev embedding theorem is a fundamental tool for proving the "regularity" properties of the solutions to differential equations. We consider this in chapter 6.

PROBLEMS FOR CHAPTER 5

5.1 a) If $u \in L^1(\mathbf{R}^n)$, let $A_u \in B(L^2(\mathbf{R}^n))$ be given by $A_u(f) = u * f$. Show $\widehat{u * f} = \hat{u}\hat{f}$.

b) Show $\mathcal{F}A_u\mathcal{F}^{-1} = M_{\hat{u}}$ where M denotes a multiplication operator.

5.2 Let $\| \ \|_{2,k}$ be the usual Sobolev norm $L^{2,k}(\mathbf{R}^n)$. Show that $\|f\| = \|(1 + \|\xi\|^k)\hat{f}\|_2$ also defines a norm on $L^{2,k}(\mathbf{R}^n)$ and that it is equivalent to $\| \ \|_{2,k}$.

5.3 If $f \in L^1(\mathbf{R}^n)$, let $V_f \subset L^1(\mathbf{R}^n)$ be the closed linear subspace generated by $\{\pi(t)f \mid t \in \mathbf{R}\}$ where π is the regular representation of $\mathbf{R}^n$ on $L^1(\mathbf{R}^n)$. Suppose that $\hat{f}(\xi_0) = 0$ for some $\xi_0 \in \hat{\mathbf{R}}^n$. Show $\hat{h}(\xi_0) = 0$ for all $h \in V_f$. Deduce that if $V_f = L^1(\mathbf{R}^n)$, then $\hat{f}$ never vanishes. (The converse is also true, being a theorem of N. Wiener.)

5.4 If $f \in C^\infty(\mathbf{R}^n)$, f is called rapidly decreasing if for every polynomial p and every α, $|p(x)D^\alpha f(x)| \to 0$ as $x \to \infty$. Let S be the space of rapidly decreasing functions (sometimes called the Schwartz space).

a) If $f \in S$, show $D^\beta f \in S$ for all β.

b) Show $S \subset L^q(\mathbf{R}^n)$ for all q, $1 \leq q \leq \infty$.

c) Show $\mathcal{F}(S) = \hat{S}$ (where $\hat{S}$ is the Schwartz space in the variable ξ).

d) Show that S is a Frechet space in such a way that $f_n \to f$ if and only if for every polynomial p and all α, $pD^\alpha f_n \to pD^\alpha f$ uniformly.

e) Show that $f_n \to f$ for the topology in (d) if and only if for every polynomial p and every α we have $pD^\alpha f_n \to pD^\alpha f$ in $L^2(\mathbf{R}^n)$.

5.5 If $f \in L^2(\mathbf{R}^n)$, show

$$\hat{f}(\xi) = \lim_{R \to \infty} \int_{\|x\| \leq R} f(x)e^{-i(x,\xi)} \, dm(x)$$

where the limit is in $L^2(\hat{\mathbf{R}}^n)$.

5.6 a) Show $C_c^\infty(\mathbf{R}^n)$ is dense in $L^{2,k}(\mathbf{R}^n)$. Hint: Let $\delta \in C_c^\infty(\mathbf{R}^n)$, $\delta \geq 0$, $\delta(x) = 1$ if $\|x\| \leq 1$. If $f \in L^{2,k}(\mathbf{R}^n)$, let $f_n(x) = \delta(x/n)f(x)$. Now use Proposition 1.1.22 (and its proof) to show $f_n \to f$ in $L^{2,k}(\mathbf{R}^n)$.

b) If $\Omega \subset \mathbf{R}^n$ is bounded, show $C_c^\infty(\Omega)$ is not dense in $L^{2,k}(\Omega)$ if k is sufficiently large. Hint: Use 5.2.6 to show $1 \notin H^k(\Omega)$.

5.7 Let Δ be the Laplace operator on $\mathbf{R}^n$, i.e., $\Delta = \sum_{i=1}^n \partial^2/\partial x_i^2$. Show that the differential operator $I - \Delta : C_c^\infty(\mathbf{R}^n) \to C_c^\infty(\mathbf{R}^n)$ extends to an isomorphism of Hilbert spaces $I - \Delta : L^{2,k}(\mathbf{R}^n) \to L^{2,k-2}(\mathbf{R}^n)$ for any $k \geq 2$.

5.8 Show $f \in L^{2,k}(\mathbf{R}^n)$ if and only if $\xi^\alpha \widehat{f} \in L^2(\widehat{\mathbf{R}}^n)$ for all $|\alpha| \leq k$.

DISTRIBUTIONS AND ELLIPTIC OPERATORS

6.1. Basic properties of distributions

We have seen in section 1.1 that one can sometimes define a notion of derivative (i.e., the weak derivative) for certain locally integrable functions that are not differentiable (or even continuous) in the usual sense. There are, however, many locally integrable functions that do not have weak derivatives in the sense of Definition 1.1.17. For example, if $f : \mathbf{R} \to \mathbf{R}$ is the characteristic function of $[0, \infty)$, then f is locally integrable, but if its weak derivative (say $d_w f / dt$) existed, it would clearly have to be 0 on $\mathbf{R} - \{0\}$ and hence 0 a.e. However, if $\varphi \in C_c^\infty(\mathbf{R})$ with $\varphi(0) \neq 0$, then

$$0 = (d_w f / dt, \varphi) = -(f, \varphi') = -\int_0^\infty \varphi'(t)\, dt = \varphi(0) \neq 0.$$

Therefore, f does not have a weak derivative. For many purposes it is useful to establish a context in which all locally integrable functions have derivatives, and it is to this theory that we now turn.

Let $\Omega \subset \mathbf{R}^n$ be open and let $C_c^\infty(\Omega)'$ be the space of all linear functions $C_c^\infty(\Omega) \to \mathbf{R}$. (For simplicity, we shall for the moment consider only $\mathbf{R}$-valued functions, and ignore considerations of the topology on $C_c^\infty(\Omega)$.) We then have a natural inclusion $L_{\text{loc}}^1(\Omega) \to C_c^\infty(\Omega)'$, $f \mapsto T_f$, where $T_f(\varphi) = (f, \varphi)$. If $f \in C^\infty(\Omega) \subset L_{\text{loc}}^1(\Omega)$, then the integration by parts formula implies that $D^\alpha f$ is uniquely determined by the relation

$$T_{D^\alpha f}(\varphi) = (-1)^{|\alpha|}(f, D^\alpha \varphi) = (-1)^{|\alpha|} T_f(D^\alpha \varphi).$$

We have observed that if $f \in L_{\text{loc}}^1(\Omega)$, it is possible that there is some $h \in L_{\text{loc}}^1(\Omega)$ satisfying $T_h(\varphi) = (-1)^{|\alpha|} T_f(D^\alpha \varphi)$ even if f is not differentiable. If so, h is uniquely determined and we defined h to be the weak α-th derivative of f. In other words, if for any $T \in C_c^\infty(\Omega)'$, we let $D^\alpha T \in C_c^\infty(\Omega)'$ be given by $(D^\alpha T)(\varphi) = (-1)^{|\alpha|} T(D^\alpha \varphi)$, then $D^\alpha : C_c^\infty(\Omega)' \to C_c^\infty(\Omega)'$ agrees with the usual D^α on $C^\infty(\Omega)$ (under the identification $C^\infty(\Omega) \subset L^1_{\text{loc}}(\Omega) \hookrightarrow C_c^\infty(\Omega)'$ above) and, under the same identification agrees with D^α_w on those $f \in L^1_{\text{loc}}(\Omega)$ possessing a weak α-th derivative. If $f \in L^1_{\text{loc}}(\Omega)$ does not possess a weak α-th derivative, we can still speak of $D^\alpha f$ as an element of $C_c^\infty(\Omega)'$. Hence $C_c^\infty(\Omega)'$ gives us a natual framework for extending differentiation to all $L^1_{\text{loc}}(\Omega)$ (and beyond). With a suitable topology on $C_c^\infty(\Omega)$, the elements of $C_c^\infty(\Omega)'$ are called generalized functions or distributions on Ω. Before turning to consideration of the relevant topology, we discuss one formal point that arises in considering complex valued functions instead of just real valued functions.

For studying complex valued functions, the pairing $\langle f, g \rangle = \int f \overline{g}$ is basic. This, of course, is not linear as a function of g, but is "conjugate–linear" or "anti–linear". That is, the map $T_f(g) = \langle f, g \rangle$ is $\mathbf{R}$-linear, but satisfies $T_f(cg) = \overline{c} T_f(g)$. For any TVS E over $\mathbf{C}$ defined by a family of seminorms, we have defined the dual space E^*. We can now define the "anti–dual" space E' to be the set of continuous $\mathbf{R}$-linear maps $\lambda : E \to \mathbf{C}$ such that $\lambda(cx) = \overline{c} \lambda(x)$ for all $c \in \mathbf{C}$. Virtually everything that one says about E^* can be carried over to E'. Perhaps the simplest way to do this is to define the conjugate of E. Namely, we define a TVS $\overline{E}$ to be the same underlying real vector space but with scalar multiplication given by $(c, x) \to \overline{c} x$, where $\overline{c} x$ is scalar multiplication in E. The seminorms on E are seminorms on $\overline{E}$, and $\overline{E}$ is therefore naturally a TVS such that the identity map $\text{id} : E \to \overline{E}$ is an isomorphism of real topological vector spaces. We then clearly have $E' = (\overline{E})^*$. (Here $=$ actually means "equals", without any identifications.) Thus, E' is itself a dual space. We also remark that the map $\lambda \to \overline{\lambda}$ defines a map $E' \to E^*$ which is an isomorphism of real topological vector spaces. If $T : E \to F$ is linear, the adjoint T^* can be defined as a linear map $F' \to E'$ as well as $F^* \to E^*$, and it clearly satisfies all the same properties.

For the remainder of this chapter we will take $(f, g) = \int f \overline{g}$ whenever this is defined. We will sometimes also denote this by $\langle f, g \rangle_0$. For each f, we set $T_f(g) = (f, g)$, so that T_f is an anti-linear map. We now turn to the topology on $C_c^\infty(\Omega)$ that we will need.

We have defined the C^∞-topology on $C^\infty(\Omega)$ in Example 1.1.10 for which $\varphi_n \to \varphi$ if and only if for each α, $D^\alpha\varphi_n \to D^\alpha\varphi$ uniformly on compact subsets of Ω. We may of course consider this as defining a topology on $C_c^\infty(\Omega) \subset C^\infty(\Omega)$. However, it is easy to see that if $f \in L^1_{\mathrm{loc}}(\Omega)$, then $T_f : C_c^\infty(\Omega) \to \mathbf{C}$ is not necessarily continuous with respect to this topology. In fact, if we simply take $\Omega = \mathbf{R}^n$ and $f(x) = 1$ for all x, then continuity of T_f in this topology would assert that $\varphi_n \in C_c^\infty(\mathbf{R}^n)$ and $D^\alpha\varphi_n \to 0$ uniformly on compact sets (for each α) implies $\int \varphi_n \to 0$. It is quite easy to construct counterexamples to this assertion by allowing φ_n to be supported increasingly far away from the origin as $n \to \infty$. On the other hand, if we have a topology on $C_c^\infty(\Omega)$ in which convergence $\varphi_n \to \varphi$ implies not only convergence in the C^∞-topology but in addition that for n sufficiently large the support of all φ_n is contained in a given compact set, then for each $f \in L^1_{\mathrm{loc}}(\Omega)$ we would have $\varphi_n \to \varphi$ implies $T_f(\varphi_n) \to T_f(\varphi)$. We now construct the required topology.

DEFINITION 6.1.1. *Let $\Omega \subset \mathbf{R}^n$ be open. For each α let $\|\ \|_\alpha$ be the seminorm on $C_c^\infty(\Omega)$ given by $\|\varphi\|_\alpha = \sup\{|D^\alpha\varphi(x)| \mid x \in \Omega\}$. For each compact $K \subset \Omega$, let $\mathcal{D}_K(\Omega) = \{\varphi \in C_c^\infty(\Omega) \mid \mathrm{supp}(\varphi) \subset K\}$, and give $\mathcal{D}_K(\Omega)$ the topology defined by the family of seminorms $\{\|\ \|_\alpha\}$ (restricted to $\mathcal{D}_K(\Omega)$.) (Thus, $\mathcal{D}_K(\Omega)$ has the C^∞-topology restricted to $\mathcal{D}_K(\Omega) \subset C^\infty(\Omega)$.) Call a seminorm $\|\ \|$ on $C_c^\infty(\Omega)$ admissible if for each compact $K \subset \Omega$, $\|\ \| : \mathcal{D}_K(\Omega) \to \mathbf{R}$ is continuous. Give $C_c^\infty(\Omega)$ the topology defined by the family of all admissible seminorms. Following common practice, when endowed with this topology we shall denote $C_c^\infty(\Omega)$ by $\mathcal{D}(\Omega)$.*

We remark that since each admissible seminorm is continuous on $\mathcal{D}_K(\Omega)$, the topology on $\mathcal{D}_K(\Omega)$ as a subspace of $\mathcal{D}(\Omega)$ is the same as the C^∞-topology on $\mathcal{D}_K(\Omega)$.

Clearly each $\|\ \|_\alpha$ is itself admissible. To see some other natural examples, suppose $h : \Omega \to \mathbf{R}$ is a locally bounded Borel function, i.e., $h|K$ is bounded for each compact $K \subset \Omega$. Then $\|\varphi\|_{\alpha,h} = \sup\{|h(x)D^\alpha\varphi(x)| \mid x \in \Omega\}$ is an admissible seminorm since h is bounded on $\mathrm{supp}(\varphi)$ for any φ. There are many other admissible seminorms. Some basic properties of $\mathcal{D}(\Omega)$ are summarized in:

PROPOSITION 6.1.2. *Let E be any TVS whose topology is given by a sufficient family of seminorms. Then:*

a) *For any sequence $\varphi_j \in \mathcal{D}(\Omega)$ we have $\varphi_j \to \varphi$ if and only if:*
   i) *There is a compact set $K \subset \Omega$ such that for all n sufficiently large, $\mathrm{supp}(\varphi_j) \subset K$; and*
   ii) *For each α, $D^\alpha \varphi_j \to D^\alpha \varphi$ uniformly.*
b) *If $T : \mathcal{D}(\Omega) \to E$ is linear, then T is continuous if and only if $T|\mathcal{D}_K(\Omega)$ is continuous for each compact $K \subset \Omega$.*
c) *If $T : \mathcal{D}(\Omega) \to E$ is linear, then T is continuous if and only if it is sequentially continuous, i.e., $\varphi_j \to \varphi$ implies $T(\varphi_j) \to T(\varphi)$.*

PROOF: a) Clearly any sequence satisfying (i) and (ii) is convergent in $\mathcal{D}(\Omega)$. Conversely, suppose $\varphi_j \to \varphi$. Since $\| \ \|_\alpha$ is admissible (ii) is clearly satisfied. To see (i), replacing φ_j by $\varphi_j - \varphi$, it suffices to assume $\varphi = 0$. If (i) is not satisfied then we can find a sequence of distinct points $x_i \in \Omega$ such that $\{x_i\} \cap K$ is finite for any compact $K \subset \Omega$ and $\varphi_{j_i}(x_i) \neq 0$ for some j_i. We can choose a locally bounded Borel function $h : \Omega \to \mathbf{R}$ such that $h(x_i) = \frac{1}{\varphi_{j_i}(x_i)}$ for each i. Then $\|\varphi_{j_i}\|_{0,h} = \sup |h\varphi_{j_i}| \geq 1$. However, $\| \ \|_{0,h}$ is admissible, so $\|\varphi_{j_i}\|_{0,h} \to 0$, which is a contradiction.

b) Suppose $T|\mathcal{D}_K(\Omega)$ is continuous for all compact $K \subset \Omega$. Let $\| \ \|$ be any continuous seminorm on E. Define $| \ |$ to be the seminorm on $\mathcal{D}(\Omega)$ given by $|\varphi| = \|T(\varphi)\|$. Since $T|\mathcal{D}_K(\Omega)$ is continuous, $| \ |$ is admissible. In particular, if $\varphi_\beta \in \mathcal{D}(\Omega)$ is a net with $\varphi_\beta \to 0$, then $\|T(\varphi_\beta)\| \to 0$. Hence T is continuous. The converse assertion is obvious.

c) This follows immediately from (a) and (b) and the fact that the topology on $\mathcal{D}_K(\Omega)$ is given by a countable family of seminorms.

DEFINITION 6.1.3. *Let $\mathcal{D}'(\Omega)$ be the anti-dual space of $\mathcal{D}(\Omega)$, i.e., $\{T : \mathcal{D}(\Omega) \to \mathbf{C} \mid T$ is anti-linear and $T(\varphi_i) \to T(\varphi)$ whenever $\varphi_i \to \varphi$ is a convergent sequence in $\mathcal{D}(\Omega)\}$. The elements of $\mathcal{D}'(\Omega)$ are called generalized functions or distributions on Ω.*

EXAMPLE 6.1.4: a) Every $f \in L^1_{\mathrm{loc}}(\Omega)$ defines an element $T_f \in \mathcal{D}'(\Omega)$ by $T_f(\varphi) = (f, \varphi)$. We thus have a natural inclusion $L^1_{\mathrm{loc}}(\Omega)$

$\hookrightarrow \mathcal{D}'(\Omega)$ and in particular an inclusion $C^\infty(\Omega) \to \mathcal{D}'(\Omega)$. (We remark that this map is is injective by the proof of Lemma 1.1.18.)
b) Every Radon measure μ on Ω (i.e., μ is finite on compact sets) defines a distribution by $T_\mu(\varphi) = \int \overline{\varphi}\, d\mu$.

We now define differentiation of distributions. We recall (Exercise 1.14) that any differential operator D on Ω has a unique formal adjoint differential operator D^* which is characterized by the condition $(D\varphi, \psi) = (\varphi, D^*\psi)$ for all $\varphi \in C^\infty(\Omega)$ and $\psi \in C_c^\infty(\Omega)$. For $D = D^\alpha$, $(D^\alpha)^* = (-1)^{|\alpha|}D^\alpha$, and for an operator of order 0, i.e., $D = M_a = $ multiplication by $a \in C^\infty(\Omega)$, we have $M_a^* = M_{\overline{a}}$. Hence if $D = \sum_{|\alpha| \leq m} a_\alpha D^\alpha$, then we have explicitly that $D^* = \sum (-1)^{|\alpha|} D^\alpha \circ M_{\overline{a}_\alpha}$, i.e.,

$$D^*\psi = \sum_{|\alpha| \leq m} (-1)^{|\alpha|} D^\alpha(\overline{a}_\alpha \psi).$$

DEFINITION 6.1.5. *If D is a differential operator and $T \in \mathcal{D}'(\Omega)$, define $D(T)$ by $D(T)(\varphi) = T(D^*\varphi)$. (That $D(T) \in \mathcal{D}'(\Omega)$ follows easily from 6.1.2.)*

The map $D : \mathcal{D}'(\Omega) \to \mathcal{D}'(\Omega)$ is thus an extension of the ordinary D on $C^\infty(\Omega) \subset \mathcal{D}'(\Omega)$ (under the inclusion of 6.1.4). Furthermore, if $f \in L^1_{\text{loc}}(\Omega)$ and f has a weak α-th derivative, $D_w^\alpha f = h \in L^1_{\text{loc}}(\Omega)$, then $D^\alpha(T_f) = T_h$, i.e., $D^\alpha f = h$ when we identify $L^1_{\text{loc}}(\Omega) \subset \mathcal{D}'(\Omega)$. In particular, by Definition 6.1.5 we may now apply a differential operator to any locally integrable function or any Radon measure and obtain a distribution.

EXAMPLE 6.1.6: a) Let $\Omega = \mathbf{R}$ and let f be the characteristic function of $[0, \infty)$. Let $df/dt \in \mathcal{D}'(\mathbf{R})$ be the distributional derivative. As we saw at the beginning of this section, for any $\varphi \in C_c^\infty(\mathbf{R})$ we have $(df/dt, \varphi) = -(f, d\varphi/dt) = \overline{\varphi}(0)$. That is, $df/dt = \delta_0$, the Dirac measure at $0 \in \mathbf{R}$.
b) We can of course take $d^n \delta_0/dt^n$ as a distribution. This is simply the distribtution $T(\varphi) = (-1)^n \delta_0(d^n\varphi/dt^n)$, i.e., $T(\varphi) = (-1)^n \overline{\varphi}^{(n)}(0)$.

Just as we can define the support of a function, we can define the support of a distribution. We first observe that by restriction we have a natural map $\mathcal{D}'(\Omega) \to \mathcal{D}'(V)$ for any open $V \subset \Omega$. We say that $T \in \mathcal{D}'(\Omega)$ is 0 on V, and write $T \mid V = 0$, if its image is 0 under this map; that is, if $T(\varphi) = 0$ for all $\varphi \in \mathcal{D}(V) \subset \mathcal{D}(\Omega)$.

LEMMA 6.1.7. *If $\{V_i\}$ is a family of open subsets of Ω and $T \in \mathcal{D}'(\Omega)$ satisfies $T \mid V_i = 0$ for all i, then T is 0 on $\bigcup V_i$.*

PROOF: Let $\varphi \in \mathcal{D}(\cup V_i)$. Then $\text{supp}(\varphi)$ is contained in a finite union of V_i, say $V_1, \ldots, V_r$. Via a partition of unity, we can write $\varphi = \sum_1^r \varphi_i$ where each $\varphi_i \in \mathcal{D}(V_i)$. Thus $T(\varphi) = \sum T(\varphi_i) = 0$.

As a consequence, we see that for each $T \in \mathcal{D}'(\Omega)$ there is a unique open subset $V_T \subset \Omega$ on which T is 0 and which is maximal with respect to this property.

DEFINITION 6.1.8. *If $T \in \mathcal{D}'(\Omega)$, the support of T (denoted by $\text{supp}(T)$) is the closed subset of Ω given by $\Omega - V_T$ where V_T is as above. Thus, $\text{supp}(T)$ is the unique smallest closed set with the property that $T(\varphi) = 0$ for all $\varphi \in \mathcal{D}(\Omega - \text{supp}(T))$.*

EXAMPLE 6.1.9: a) If $x \in \Omega$, let δ_x be the Dirac measure at x. Then $\text{supp}(\delta_x) = \{x\}$. Similarly, $\text{supp}(D^\alpha \delta_x) = \{x\}$, and in fact $\text{supp}(D\delta_x) = \{x\}$ for any non-zero differential operator with constant coefficients. One can show, although we shall not do so, that any distribution with support $\{x\}$ is of this form.
b) If $f \in C^\infty(\Omega)$, then $\text{supp}(f) = \text{supp}(T_f)$; i.e., the support of f as a distribution is the same as the usual notion of support.
c) For any $T \in \mathcal{D}'(\Omega)$ and any differential operator D, $\text{supp}\big(D(T)\big) \subset \text{supp}(T)$.
d) If D is a differential operator and all coefficients have support contained in a closed set A, then $\text{supp}\big(D(T)\big) \subset A$ for any T. In particular, if $\psi \in C^\infty(\Omega)$, then $\text{supp}(\psi T) \subset \text{supp}(\psi)$.
e) $\text{supp}(T + S) \subset \text{supp}(T) \cup \text{supp}(S)$.

DEFINITION 6.1.10. *Let $\mathcal{D}'_c(\Omega)$ be the subset of $\mathcal{D}'(\Omega)$ consisting of distributions whose support is a compact subset of Ω.*

We clearly have $C_c^\infty(\Omega) \subset \mathcal{D}'_c(\Omega)$ under the standard inclusion $C^\infty(\Omega) \subset \mathcal{D}'(\Omega)$.

PROPOSITION 6.1.11. *For $T \in \mathcal{D}'(\Omega)$, the map $T : C_c^\infty(\Omega) \to \mathbb{C}$ is continuous in the C^∞-topology if and only if $T \in \mathcal{D}_c'(\Omega)$. In fact, $\mathcal{D}_c'(\Omega)$ is exactly the anti-dual space of $C_c^\infty(\Omega)$ where the latter has the C^∞-topology.*

PROOF: If $T \in \mathcal{D}_c'(\Omega)$, let $\psi \in C_c^\infty(\Omega)$ with $\psi = 1$ on a neighborhood of supp(T). If $\varphi \in C_c^\infty(\Omega)$, then $T(\psi\varphi) = T(\varphi)$, because $\psi\varphi - \varphi$ has support in $\Omega - \text{supp}(T)$, and hence $T(\psi\varphi - \varphi) = 0$. Now suppose $\varphi, \varphi_n \in C_c^\infty(\Omega)$ and $\varphi_n \to \varphi$ in the C^∞-topology. Then $\psi\varphi_n \to \psi\varphi$ in $\mathcal{D}(\Omega)$, so

$$T(\varphi_n) = T(\psi\varphi_n) \to T(\psi\varphi) = T(\varphi),$$

showing continuity of T in the C^∞-topology. To see the converse assertion, suppose supp(T) is not compact. Choose $x_n \in \text{supp}(T)$ such that x_n has no subsequence convergent in Ω. We can choose disjoint open sets $V_n \subset \Omega$ with $x_n \in V_n$ and such that each compact $K \subset \Omega$ intersects only finitely many V_n. Since $x_n \in \text{supp}(T)$, we can choose $\varphi_n \in C_c^\infty(V_n)$ such that $T(\varphi_n) \neq 0$. For any compact $K \subset \Omega$, we have for sufficiently large n that $\varphi_n | K = 0$, and in particular, for any constants c_n, we have $c_n\varphi_n \to 0$ in the C^∞-topology. However, letting $\frac{1}{c_n} = |T(\varphi_n)|$, we have $|T(c_n\varphi_n)| = 1$ for all n, showing that T is not continuous. Finally, to see the last assertion of the proposition, we observe that any $\lambda \in C_c^\infty(\Omega)'$ is a distribution, since the identity map $\mathcal{D}(\Omega) \to C_c^\infty(\Omega)$ is continuous.

Proposition 6.1.11 identifies a natural class of distributions by identifying them with those distributions that are continuous with respect to a topology smaller than that on $\mathcal{D}(\Omega)$ given by Definition 6.1.1. There are a number of other such natural classes of distributions, some of which will be discussed in section 6.2. Another such class, the "tempered" distributions, is discussed in exercise 6.10.

The distributions with compact support can be nicely described in terms of derivatives of measures.

PROPOSITION 6.1.12. *Suppose $T \in \mathcal{D}_c'(\Omega)$. Then there is an integer k and a family of (finite) complex measures μ_α, one for each $|\alpha| \leq k$, such that*

$$T = \sum_{|\alpha| \leq k} D^\alpha(\mu_\alpha).$$

PROOF: The C^∞-topology on $C_c^\infty(\Omega)$ is given by the family of seminorms $\| \ \|_{\alpha,K}$ where $K \subset \Omega$ is compact. By 6.1.11, $T^{-1}(\{|z| < 1\})$ is open in this topology, so there is a finite collection $\alpha_1, \ldots, \alpha_r$, compact sets $K_1, \ldots, K_r \subset \Omega$, and positive $\varepsilon_1, \ldots, \varepsilon_r$ such that

$$T^{-1}(\{|z| < 1\}) \supset \bigcap_1^r \{\varphi \mid \|\varphi\|_{\alpha_i,K_i} < \varepsilon_i\}.$$

Let $k = \max |\alpha_i|$, $K = \cup K_i$, and for each $|\alpha| \leq k$, let $E_\alpha = C(K)$. Define

$$\theta : \mathcal{D}(\Omega) \to E = \sum_{|\alpha| \leq k}^{\oplus} E_\alpha$$

by $\theta(\varphi) = (\varphi_\alpha)$ where $\varphi_\alpha = (D^\alpha \varphi)|K$. Then θ is linear and continuous, and T factors to a map (which we still denote by T) on $\theta(\mathcal{D}(\Omega))$ (cf. the proof of 1.1.30). Furthermore, by the choice of α_i, K, if $\theta(\varphi_j) \to 0$ then $T(\varphi_j) \to 0$. Thus, $T : \theta(\mathcal{D}(\Omega)) \to \mathbb{C}$ is continuous. By the Hahn-Banach theorem, T extends to an anti-linear map $T' : E \to \mathbb{C}$, and by the Riesz representation theorem (A.20) there are complex measures μ_α on K, $|\alpha| \leq k$, such that

$$T'((f_\alpha)) = \sum_{|\alpha| \leq k} (-1)^{|\alpha|} \int \overline{f}_\alpha \, d\mu_\alpha.$$

Therefore, for $\varphi \in \mathcal{D}(\Omega)$

$$T(\varphi) = T'(\theta(\varphi)) = \sum_{|\alpha| \leq k} (-1)^{|\alpha|} \int D^\alpha \overline{\varphi} \, d\mu_\alpha.$$

Hence, $T = \sum_{|\alpha| \leq k} D^\alpha(\mu_\alpha)$ as required.

REMARK: We note that the proposition implies that for any $T \in \mathcal{D}_c'(\Omega)$, there is some k such that $T : C_c^\infty(\Omega) \to \mathbb{C}$ is continuous with respect to the $BC^k(\Omega)$-norm on $C_c^\infty(\Omega)$.

Here is another useful remark about $\mathcal{D}_c'(\Omega)$.

PROPOSITION 6.1.13. *Suppose $T \in \mathcal{D}'_c(\Omega)$. Then T can be extended in a unique way to an element T' of $\mathcal{D}'_c(\mathbf{R}^n)$ with $\operatorname{supp}(T') = \operatorname{supp}(T)$. If D is any differential operator, then $D(T') = D(T)'$. (As is customary, in the sequel we shall usually denote this extension simply by T.)*

PROOF: Let $\psi \in C^\infty_c(\Omega) \subset C^\infty_c(\mathbf{R}^n)$ with $\psi = 1$ on a neighborhood of $\operatorname{supp}(T)$. As in the proof of 6.1.11, we have $T(\varphi) = T(\psi\varphi)$ for all $\varphi \in \mathcal{D}(\Omega)$. For any $\varphi \in C^\infty_c(\mathbf{R}^n)$, define $T'(\varphi) = T(\psi\varphi)$. Then one easily verifies all the stated properties.

6.2. Distributions and Sobolev spaces

We now consider those distributions that are continuous with respect to the Sobolev norms on $C^\infty_c(\Omega)$. We shall only be considering Sobolev spaces of the form $L^{p,k}(\Omega)$ for $p = 2$. We let $H^k(\Omega)$ ($k \geq 0$) be the closure of $C^\infty_c(\Omega)$ in $L^{2,k}(\Omega)$, and let $\| \ \|_k$ denote the norm on $H^k(\Omega)$. Although we shall not be using these facts, we recall that $H^k(\mathbf{R}^n) = L^{2,k}(\mathbf{R}^n)$, but if Ω is bounded this is no longer necessarily true (cf. exercise 5.6).

DEFINITION 6.2.1. *For $k > 0$, let $H^{-k}(\Omega) = \{T \in \mathcal{D}'(\Omega) \mid T$ is continuous with respect to $\| \ \|_k$ on $C^\infty_c(\Omega)\}$.*

Each element $T \in H^{-k}(\Omega)$ thus extends uniquely to an element $T \in H^k(\Omega)'$. That is, we have a natural map $H^{-k}(\Omega) \to H^k(\Omega)'$. Since $C^\infty_c(\Omega)$ is dense in $H^k(\Omega)$, this is injective by definition, and since $\mathcal{D}(\Omega) \to \left(C^\infty_c(\Omega), \| \ \|_k\right)$ is continuous, the map is a bijection. We then give $H^{-k}(\Omega)$ the norm defined by this identification with the Hilbert space $H^k(\Omega)'$, and denote this norm by $\| \ \|_{-k}$. Thus, if $T \in H^{-k}(\Omega)$, $\|T\|_{-k} = \sup\{|T(f)| \mid f \in H^k(\Omega), \|f\|_k \leq 1\}$. We have observed (cf. Remark 1.1.9) that one may choose a variety of equivalent norms on $H^k(\Omega)$ for $k \geq 0$. The space $H^{-k}(\Omega)$ is independent of the choice but of course the particular norm on $H^{-k}(\Omega)$ depends upon that on $H^k(\Omega)$.

LEMMA 6.2.2. a) $H^0(\Omega) \subset H^{-k}(\Omega)$ for any $k \geq 0$.
b) $C^\infty_c(\Omega)$ is dense in $H^{-k}(\Omega)$.

(In both (a) and (b), we make the standard identification $L^1_{\text{loc}}(\Omega) \hookrightarrow \mathcal{D}'(\Omega)$.)

PROOF: (a) is clear. To see (b), since $H^{-k}(\Omega) \cong H^k(\Omega)'$ it suffices to see that if $f \in H^k(\Omega)$ and $T_\varphi(f) = 0$ for all $\varphi \in C_c^\infty(\Omega)$, then $f = 0$. But $T_\varphi(f) = (f, \varphi)$ and since $f \in L^2$, this follows from Corollary B.6.

We now have a sequence of Hilbert spaces $H^k(\Omega)$ indexed by $k \in \mathbf{Z}$, each of which can be viewed as the completion of $C_c^\infty(\Omega)$ with respect to a norm $\| \ \|_k$. We have an inclusion $H^k(\Omega) \subset H^\ell(\Omega)$ for any $k, \ell \in \mathbf{Z}$ with $k \geq \ell$, and this inclusion extends the identity $C_c^\infty(\Omega) \rightarrow C_c^\infty(\Omega)$. Furthermore, by 6.2.2(b), it is clear that even for $k \geq 0$, we may view $H^k(\Omega) = \{T \in \mathcal{D}'(\Omega) \mid T \text{ is continuous with respect to } \| \ \|_{-k}\}$, and that with this identification of $H^k(\Omega) \subset L^1_{\text{loc}}(\Omega)$ as distributions, we have $H^k(\Omega) \cong H^{-k}(\Omega)'$.

PROPOSITION 6.2.3. a) If $T \in \mathcal{D}'_c(\Omega)$, then $T \in H^{-k}(\Omega)$ for k sufficiently large.
b) If $T \in H^{-k}(\Omega)$, then for $|\alpha| \leq k$ there are $f_\alpha \in L^2(\Omega)$, such that $T = \sum_{|\alpha| \leq k} D^\alpha f_\alpha$.
c) If $T \in \mathcal{D}'_c(\Omega) \cap H^{-k}(\Omega)$ then $T \in H^{-k}(\mathbf{R}^n)$ where we extend T to an element of $\mathcal{D}'_c(\mathbf{R}^n)$ by 6.1.13.

PROOF: a) By the remark following Proposition 6.1.12, there is some $r > 0$ such that T is continuous with respect to the $BC^r(\Omega)$-topology on $C_c^\infty(\Omega)$. By the Sobolev embedding theorem (in the form of Corollary 5.2.6), for k sufficiently large, $H^k(\Omega) \rightarrow BC^r(\Omega)$ is continuous. Thus, T is also continuous with respect to $\| \ \|_k$, and hence lies in $H^{-k}(\Omega)$.
b) Since T defines and element of $H^k(\Omega)'$, there is some $h \in H^k(\Omega)$ such that

$$T(\varphi) = \langle h, \varphi \rangle_k = \sum_{|\alpha| \leq k} (D_w^\alpha h, D^\alpha \varphi).$$

Thus, letting $f_\alpha = (-1)^{|\alpha|} D_w^\alpha h$, we have

$$T(\varphi) = \sum_{|\alpha| \leq k} T_{f_\alpha}((D^\alpha)^* \varphi),$$

so $T = \sum D^\alpha f_\alpha$.

c) Choose $\psi \in C_c^\infty(\Omega)$ with $\psi = 1$ on a neighborhood of supp(T). If $\varphi_n, \varphi \in C_c^\infty(\mathbf{R}^n)$ and $\varphi_n \to \varphi$ in $\|\ \|_k$, then $\psi\varphi_n \to \psi\varphi$ in $\|\ \|_k$ as well. Thus,

$$T(\varphi_n) = T(\psi\varphi_n) \to T(\psi\varphi) = T(\varphi).$$

We remark that 6.2.3 yields another type of structural feature of elements of $\mathcal{D}_c'(\Omega)$ in a similar spirit to 6.1.12.

Now suppose D is a differential operator on Ω of order ≤ 1, such that all coefficients are in $BC^\infty(\Omega)$. We have already remarked (Example 1.2.12) that D extends to a bounded operator $L^{2,k}(\Omega) \to L^{2,k-1}(\Omega)$ for any $k \geq 1$, and in particular to a bounded operator $H^k(\Omega) \to H^{k-1}(\Omega)$ for $k \geq 1$. Suppose now $k \geq 0$. If $T \in H^{-k}(\Omega)$, we have defined $D(T) \in \mathcal{D}'(\Omega)$. For $\varphi \in C_c^\infty(\Omega)$, we have $\big(D(T)\big)(\varphi) = T\big(D^*(\varphi)\big)$, so that

$$|D(T)\varphi| \leq \|T\|_{-k}\|D^*(\varphi)\|_k \leq c\|T\|_{-k}\|\varphi\|_{k+1}$$

where c is the norm of $D^* : H^{k+1}(\Omega) \to H^k(\Omega)$. It follows that $D(T) \in H^{-(k+1)}(\Omega)$ and that $D : H^{-k}(\Omega) \to H^{-k-1}(\Omega)$ is bounded with norm at most c. In other words, for any integer k, $D : C_c^\infty(\Omega) \to C_c^\infty(\Omega)$ extends to a bounded operator $H^k(\Omega) \to H^{k-1}(\Omega)$. Furthermore, it is essentially immediate from the definitions that the adjoint of this map, which is a bounded map $H^{k-1}(\Omega)' \to H^k(\Omega)'$, is, under the identification of $H^k(\Omega)'$ with $H^{-k}(\Omega)$ simply the map $D^* : H^{-k+1}(\Omega) \to H^{-k}(\Omega)$ defined by the formal adjoint differential operator (which we recall is also of order ≤ 1 with coefficients in $BC^\infty(\Omega)$.) Since any differential operator of order m with coefficients in $BC^\infty(\Omega)$ is a finite linear combination of compositions of at most m operators of order ≤ 1 with $BC^\infty(\Omega)$ coefficients, we deduce:

PROPOSITION 6.2.4. *Let D be a differential operator of order $\leq m$ on Ω with all coefficients in $BC^\infty(\Omega)$. Then for each $k \in \mathbf{Z}$, $D : C_c^\infty(\Omega) \to C_c^\infty(\Omega)$ extends to a bounded operator $D : H^k(\Omega) \to H^{k-m}(\Omega)$. If $k \leq 0$, this operator agrees with the action of D on* *distributions given in 6.1.5. The adjoint operator $H^{k-m}(\Omega)' \rightarrow H^k(\Omega)'$, when identified as a map $H^{-k+m}(\Omega) \rightarrow H^{-k}(\Omega)$, is the operator defined by the formal adjoint D^*.*

The Laplace operator plays a special role in the context of 6.2.4, which we shall see in 6.2.9 below. We now describe this and at the same time indicate the relation of H^k to the Fourier transform.

We recall that the Laplace operator is the differential operator $\Delta = \sum_1^n \partial_i^2$ (where we continue to write $\partial_i = \partial/\partial x_i$). We recall that the formal adjoint of ∂_i is given by $\partial_i^* = -\partial_i$, and hence $\Delta = -\sum \partial_i^* \partial_i$. It follows that Δ is formally self-adjoint, i.e., $\Delta^* = \Delta$. We recall once again that we have some choice in the norm on $H^k(\Omega)$ (which then determines the norm on $H^{-k}(\Omega)$). We shall now fix a choice of these norms which will be very convenient in that they are very well behaved with respect to Δ. For each positive integer k, we have

$$(I - \Delta)^k = (I + \sum_1^n \partial_i^* \partial_i)^k = \sum_{|\alpha| \leq k} c_\alpha (D^\alpha)^* D^\alpha$$

where $c_\alpha \in \mathbf{Z}$ are strictly positive integers. Here c_α is simply the coefficient of ξ^α in the polynomial $p(\xi) = (1 + \sum_1^n \xi_i)^k$. Thus, if we define the inner product $\langle \ , \ \rangle_k$ on $H^k(\Omega)$ by

$$\langle \varphi, \psi \rangle_k = \sum_{|\alpha| \leq k} c_\alpha (D^\alpha \varphi, D^\alpha \psi),$$

then we clearly have an inner product on $H^k(\Omega)$ whose norm is equivalent to a standard one. Henceforth, we shall always assume $\| \ \|_k$ (and hence $\| \ \|_{-k}$) are taken with respect to this choice of inner product. We then easily have the following facts, which we shall improve upon shortly for $\Omega = \mathbf{R}^n$. (Cf. exercise 6.15.)

PROPOSITION 6.2.5. a)   *For any positive integer m, the inner product $\langle \varphi, \psi \rangle_m$ on $H^m(\Omega)$ is given by $\langle \varphi, \psi \rangle_m = ((I - \Delta)^m \varphi, \psi)$.*
b) *For each m, $(I - \Delta)^m : H^{2m}(\Omega) \rightarrow H^0(\Omega)$ is an isometry.*

c) *For each* m, $(I - \Delta)^m : H^m(\Omega) \to H^{-m}(\Omega)$ *is an isometric isomorphism.*

PROOF: a) is immediate from the definition.

b) Observe that

$$\|(I - \Delta)^m \varphi\|_0^2 = ((I - \Delta)^m \varphi, (I - \Delta)^m \varphi) = ((I - \Delta)^{2m} \varphi, \varphi)$$

since $\Delta = \Delta^*$. By a), this equals $\|\varphi\|_{2m}$.

c) From $\langle \varphi, \psi \rangle_m = ((I - \Delta)^m \varphi, \psi)$ for $\varphi, \psi \in C_c^\infty(\Omega)$, we see that

$$\|(I - \Delta)^m \varphi\|_{-m} = \sup\{|\langle \varphi, \psi \rangle_m| \,\big|\, \|\psi\|_m \leq 1\} = \|\varphi\|_m$$

Thus, $(I - \Delta)^m$ is an isometry. To see it is an isomorphism, it suffices to see that the adjoint

$$\left((I - \Delta)^m\right)^* : H^{-m}(\Omega)' \to H^m(\Omega)'$$

is bounded below (cf. Lemma 4.1.6). However, by 6.2.4 and the fact that $\Delta = \Delta^*$, this can be identified with $(I - \Delta)^m : H^m(\Omega) \to H^{-m}(\Omega)$, which we have just seen is bounded below.

We have seen in Chapter 5 the utility of the Fourier transform for understanding Sobolev spaces, and it is useful to consider the spaces H^{-k} in this context. We shall now take $\Omega = \mathbf{R}^n$. For functions on $\widehat{\mathbf{R}}^n$ we shall only be considering the L^2-norm and the usual inner product, and we shall always take $\langle \, , \rangle$ and $\| \, \|$ to be these standard ones when applied to functions defined on $\widehat{\mathbf{R}}^n$.

We recall from Proposition 5.2.1 that if $f \in L^{2,k}(\mathbf{R}^n)$, then for any $|\alpha| \leq k$ we have

$$(\widehat{D_w^\alpha f})(\xi) = (-i)^{|\alpha|} \xi^\alpha \widehat{f}(\xi).$$

Therefore if $f \in C_c^\infty(\mathbf{R}^n)$,

$$((1 - \Delta)^k f)^\wedge = (1 + |\xi|^2)^k \widehat{f}(\xi).$$

By 6.2.5(a) we have, for $f, g \in C_c^\infty(\mathbf{R}^n)$ that

$$\langle f, g \rangle_k = ((1 - \Delta)^k f, g) = \langle (1 + |\xi|^2)^k \widehat{f}, \widehat{g} \rangle$$

via the Plancherel theorem. In particular,

$$\|f\|_k = \|(1 + |\xi|^2)^{k/2} \widehat{f}\|.$$

We can thus alternatively view $H^k(\mathbf{R}^n)$ ($k \geq 0$) as the completion of $C_c^\infty(\mathbf{R}^n)$ with respect to the norm $\varphi \mapsto \|(1 + |\xi|^2)^{k/2} \widehat{\varphi}\|$.

DEFINTION 6.2.6. *For each $k \in \mathbf{R}$, define $\| \ \|_k^{\wedge}$ on $C_c^{\infty}(\mathbf{R}^n)$ by*

$$\|\varphi\|_k^{\wedge} = \|(1 + |\xi|^2)^{k/2}\widehat{\varphi}\|,$$

and let H_k be the completion of $C_c^{\infty}(\mathbf{R}^n)$ with respect to this norm. This is a Hilbert space with the inner product on $C_c^{\infty}(\mathbf{R}^n)$ given by

$$\langle \varphi, \psi \rangle_k^{\wedge} = \langle (1 + |\xi|^2)^k \widehat{\varphi}, \widehat{\psi} \rangle.$$

As remarked above, for a non-negative integer k we have $H^k(\mathbf{R}^n) = H_k$ and $\| \ \|_k = \| \ \|_k^{\wedge}$.

The following simple observation is useful.

LEMMA 6.2.7. *Let $k \geq 0$. Then*

$$\{(1 + |\xi|^2)^{k/2}\widehat{\varphi} \mid \varphi \in H^k(\mathbf{R}^n)\} = L^2(\widehat{\mathbf{R}}^n),$$

and for $\varphi \in H^k(\mathbf{R}^n)$,

$$\|\varphi\|_k = \|(1 + |\xi|^2)^{k/2}\widehat{\varphi}\|.$$

PROOF: If $\varphi \in H^k(\mathbf{R}^n)$, then by Proposition 5.2.1 $\xi^{\alpha}\widehat{\varphi} \in L^2(\mathbf{R}^n)$ for $|\alpha| \leq k$. Therefore $\xi^{2\alpha}|\widehat{\varphi}|^2 \in L^1(\mathbf{R}^n)$, and hence $(1+|\xi|^2)^k|\widehat{\varphi}|^2 \in L^1(\mathbf{R}^n)$. Therefore $(1+|\xi|^2)^{k/2}\widehat{\varphi} \in L^2(\mathbf{R}^n)$, and $\|(1+|\xi|^2)^{k/2}\widehat{\varphi}\|^2 = \|\varphi\|_k^2$ by 5.2.1 and the Plancherel theorem. Conversely if $h \in L^2(\widehat{\mathbf{R}}^n)$, let $h_1 = (1 + |\xi|^2)^{-k/2}h$, so that $h_1 \in L^2(\widehat{\mathbf{R}}^n)$. Let $\psi = \widetilde{h}_1$ (the inverse Fourier transform). Then $\psi \in L^2(\mathbf{R}^n)$ and $(1 + |\xi|^2)^{k/2}\widehat{\psi} = h$. Thus $\psi \in H^k(\mathbf{R}^n)$, and the set in question contains h.

Fix k to be a non-negative integer. For $\varphi, \psi \in C_c^{\infty}(\mathbf{R}^n)$, we have

$$\begin{aligned}
(\varphi, \psi) = \langle \widehat{\varphi}, \widehat{\psi} \rangle &= \int [(1 + |\xi|^2)^{-k/2}\widehat{\varphi}][(1 + |\xi|^2)^{k/2}\overline{\widehat{\psi}}] \\
&\leq \|(1 + |\xi|^2)^{-k/2}\widehat{\varphi}\|\|(1 + |\xi|^2)^{k/2}\overline{\widehat{\psi}}\| \\
&= \|\varphi\|_{-k}^{\wedge}\|\psi\|_k^{\wedge}.
\end{aligned}$$

Thus, the map $\varphi \mapsto (\varphi, \cdot)$ extends to a bounded linear map $F : H_{-k} \to H_k'$. This enables us to give the following description of $H^{-k}(\mathbf{R}^n)$ in terms of the Fourier transform.

PROPOSITION 6.2.8. *The map* $F : H_{-k} \to (H_k)' = H^{-k}(\mathbf{R}^n)$ *is an isometric isomorphism. Thus we can (and henceforth shall) identify* $(H_{-k}, \| \ \|^{\wedge}_{-k})$ *with* $(H^{-k}(\mathbf{R}^n), \| \ \|_{-k})$.

PROOF: From $(\varphi, \psi) \leq \|\varphi\|^{\wedge}_{-k}\|\psi\|^{\wedge}_k$ and $\|\psi\|^{\wedge}_k = \|\psi\|_k$, we deduce that $\|F(\varphi)\|_{-k} \leq \|\varphi\|^{\wedge}_{-k}$. In fact, we have

$$\|F(\varphi)\|_{-k} = \sup\{|(\varphi, \psi)| \, | \, \psi \in H^k(\mathbf{R}^n), \|\psi\|_k \leq 1\}$$
$$= \sup\{|\langle(1 + |\xi|^2)^{-k/2}\widehat{\varphi}, (1 + |\xi|^2)^{k/2}\widehat{\psi}\rangle| \, | \, \|\psi\|_k \leq 1\}.$$

By Lemma 6.2.7, this

$$= \sup\{|\langle(1 + |\xi|^2)^{-k/2}\widehat{\varphi}, h\rangle| \, | \, h \in L^2(\widehat{\mathbf{R}}^n), \|h\| \leq 1\}$$
$$= \|(1 + |\xi|^2)^{-k/2}\widehat{\varphi}\|$$
$$= \|\varphi\|^{\wedge}_{-k}.$$

Thus F is an isometry, and to see it is an isomorphism it suffices to see the image is dense (4.1.6). However, it contains $C^{\infty}_c(\mathbf{R}^n) \subset (H^k(\mathbf{R}^n))'$, so the result follows from Lemma 6.2.2.

We now give a sharper version of 6.2.5 for $\Omega = \mathbf{R}^n$.

PROPOSITION 6.2.9. *For each* $k \in \mathbf{Z}$ *and each integer* $m \geq 0$, *we have:*
a) $(I - \Delta)^m : H^{2m+k}(\mathbf{R}^n) \to H^k(\mathbf{R}^n)$ *is an isometric isomorphism;*
b) $\langle(I - \Delta)^m\varphi, \psi\rangle_k = \langle\varphi, \psi\rangle_{k+m}$

PROOF: a) Case 1: $k \geq 0$. Under the Fourier transform $(I - \Delta)^m$ corresponds to multiplication by $(1 + |\xi|^2)^m$, and from the definition of the norms on H_k, it is an isometry. To see it is an isomorphism, we argue exactly as in the proof of 6.2.7. Namely, if $\psi \in H^k(\mathbf{R}^n)$, then $\widehat{\psi}, (1 + |\xi|^2)^{k/2}\widehat{\psi} \in L^2(\widehat{\mathbf{R}}^n)$, and $h = (1 + |\xi|^2)^{-m}\widehat{\psi} \in L^2(\mathbf{R}^n)$. Let $f = \widetilde{h}$, so that $f \in L^2(\mathbf{R}^n)$ and $(1 + |\xi|^2)^{(2m+k)/2}\widehat{f} \in L^2(\widehat{\mathbf{R}}^n)$. Thus, $f \in H^{2m+k}(\mathbf{R}^n)$ and $((I - \Delta)^m f)^{\wedge} = (1 + |\xi|^2)^m\widehat{f} = \widehat{\psi}$. Thus, $(I - \Delta)^m f = \psi$.
Case 2: $2m + k \leq 0$. This follows from case 1 since the map in question is simply the adjoint of one considered in case 1.

Case 3: If $2m + k > 0$ and $k < 0$, then $(I - \Delta)^m$ will be the composition of powers of $(I - \Delta)$ from case 1 and case 2, and if k is odd, the case of 6.2.5(c).

b) is immediate from Definition 6.2.6 and the identification given by 6.2.8.

We remark that 6.2.9(a) is not true if $\mathbf{R}^n$ is replaced by a bounded open set. See exercise 6.15.

We have defined the local L^p spaces in chapter 1. (See Example 1.1.7.) We recall that for a measurable $f : \Omega \to \mathbf{C}$, we have $f \in L^p_{\mathrm{loc}}(\Omega)$ means that $f|K \in L^p(K)$ for every compact $K \subset \Omega$. Equivalently, $\varphi f \in L^p(\Omega)$ for every $\varphi \in C^\infty_c(\Omega)$. We now define the local Sobolev spaces.

DEFINITION 6.2.10. *For $\Omega \subset \mathbf{R}^n$ open, and any $k \in \mathbf{Z}$, define*

$$H^k_{\mathrm{loc}}(\Omega) = \{T \in \mathcal{D}'(\Omega) \mid \psi T \in H^k(\Omega) \text{ for every } \psi \in C^\infty_c(\Omega)\}.$$

We clearly have $H^k(\Omega) \subset H^k_{\mathrm{loc}}(\Omega)$ and $H^0_{\mathrm{loc}}(\Omega) = L^2_{\mathrm{loc}}(\Omega)$. Here are some basic properties.

PROPOSITION 6.2.11. a) *If $T \in \mathcal{D}'_c(\Omega) \cap H^k_{\mathrm{loc}}(\Omega)$, then $T \in H^k(\Omega)$.*
b) *Suppose D is a differential operator of order m on Ω with coefficients in $BC^\infty(\Omega)$. Then $D : H^k_{\mathrm{loc}}(\Omega) \to H^{k-m}_{\mathrm{loc}}(\Omega)$*
c) *If D is a differentiable operator of order m on Ω with coefficients in $C^\infty_c(\Omega)$, then*

$$D : H^k_{\mathrm{loc}}(\Omega) \to H^{k-m}(\Omega).$$

PROOF: a) Let $\psi \in C^\infty_c(\Omega)$ with $\psi = 1$ on a neighborhood of supp(T). If $\varphi_n \to \varphi_0$ in $C^\infty_c(\Omega)$ with $\| \ \|_{-k}$, then $\psi\varphi_n \to \psi\varphi_0$ in $\| \ \|_{-k}$. Thus

$$T(\psi\varphi_n) = (\psi \cdot T)(\varphi_n) \to (\psi T)(\varphi_0) = T(\psi\varphi_0).$$

However, for all $n \geq 0$, $T(\psi\varphi_n - \varphi_n) = 0$ since $\psi\varphi_n - \varphi_n$ vanishes on a neighborhood of supp(T). Thus $T(\varphi_n) \to T(\varphi_0)$, showing $T \in H^k(\Omega)$.

b) By induction, it suffices to consider the cases $D = \partial_i$ for some i, or $D(\varphi) = a\varphi$ where $a \in BC^\infty(\Omega)$ (i.e., order $D = 0$). Let $\varphi_n \to \varphi_0$ as in (a), and let $\psi \in C_c^\infty(\Omega)$. Let $T \in H_{\text{loc}}^k(\Omega)$. Then

$$\Big(\psi(D(T))\Big)(\varphi_n) = T\Big(D^*(\psi\varphi_n)\Big).$$

If $D(\varphi) = a\varphi$, then

$$\Big(\psi D(T)\Big)(\varphi_n) = T(\overline{a}\psi\varphi_n) = (\psi T)(\overline{a}\varphi_n)$$
$$\longrightarrow (\psi T)(\overline{a}\varphi_0) = \Big(\psi D(T)\Big)(\varphi_0).$$

For $D = \partial_i$, and $\varphi_n \to \varphi_0$ in $\| \ \|_{-k+1}$ we have

$$\Big(\psi(D(T))\Big)(\varphi_n) = -T\Big(\partial_i(\psi\varphi_n)\Big)$$
$$= -T\Big(\psi\partial_i(\varphi_n)\Big) - T\Big(\partial_i(\psi)\varphi_n\Big)$$
$$= -(\psi T)\Big(\partial_i(\varphi_n)\Big) - \Big(\partial_i(\psi)T\Big)(\varphi_n).$$

If $T \in H_{\text{loc}}^k(\Omega)$, this converges to

$$-(\psi T)\Big(\partial_i(\varphi_0)\Big) - \Big(\partial_i(\psi)T\Big)(\varphi_0) = -T\Big(\psi\partial_i(\varphi_0) + \partial_i(\psi)\varphi_0\Big)$$
$$= \Big(\psi D(T)\Big)(\varphi_0).$$

c) Since the coefficients of D have compact support, so does $D(T)$ by 6.1.9(c) and hence (c) follows from (a) and (b).

The Sobolev embedding theorem (5.2.5) immediately implies:

COROLLARY 6.2.12. If $T \in \mathcal{D}'(\Omega)$ and $T \in H_{\text{loc}}^r(\Omega)$ for all r, then $T \in C^\infty(\Omega)$.

6.3. Regularity for elliptic operators

In this section, we consider the following problem concerning differential operators. Suppose $DT = u$ where D is a differential operator and T and u are distributions. If u has certain regularity properties, the question arises as to what properties T must have. For example, if u is a smooth function, must T be a smooth function? One of the ways (but not the only way) in which such a question arises is in trying to solve the differential equation $Df = u$. Here D and u are known and we are trying to find f. There are many approaches which enable one (under suitable hypotheses) to find a distribution f satisfying the equation, and hence the question as to when one can deduce that f is a smooth function and therefore a solution to the equation in the classical sense. We shall not consider the question of existence of distribution solutions here, but many of the techniques of this chapter are relevant. Instead, we shall focus on proving a basic regularity result for a natural class of differential operators.

EXAMPLE 6.3.1: a) On $\mathbf{R}^2$ consider the equation $\partial_1 f = u$. If $u = 0$, and we let $f(x, y) = h(y)$ for any $h \in L^1_{\text{loc}}(\mathbf{R})$, then clearly $\partial_1 f = 0$. Here u is smooth but f is not necessarily continuous.
b) On $\mathbf{R}^2$, let $D = \partial_1 + \partial_2$. This operator reduces to ∂_1 after a linear change of coordinates. Thus, if $f(x, y) = h(x - y)$ where $h \in L^1_{\text{loc}}(\mathbf{R})$, then $Df = 0$.

On the other hand, we now indicate a first regularity result for the Laplace operator.

PROPOSITION 6.3.2. *Suppose $\Delta f = u$ where $u \in H^r(\mathbf{R}^n)$ for all r (and in particular is smooth). If $f \in H^k(\mathbf{R}^n)$ for some k (e.g., $f \in L^2(\mathbf{R}^n)$), then $f \in C^\infty(\mathbf{R}^n)$.*

PROOF: The equation $\Delta f = u$ is an equation in $H^{k-2}(\mathbf{R}^n)$, so we have $(I - \Delta)f = f - u$ in $H^{k-2}(\mathbf{R}^n)$. However, $f - u \in H^k(\mathbf{R}^n)$, so there is a unique $g \in H^{k+2}(\mathbf{R}^n)$ such that $(I - \Delta)g = f - u$ (Proposition 6.2.9). Since this is also an equation in $H^{k-2}(\mathbf{R}^n)$, we have $f = g$, i.e., $f \in H^{k+2}(\mathbf{R}^n)$. Continuing by induction, we deduce that $f \in H^r(\mathbf{R}^n)$ for all r and by the Sobolev embedding theorem, $f \in C^\infty(\mathbf{R}^n)$.

While Proposition 6.3.2 is very suggestive, it raises (at least) two questions immediately. First is the question of localizing this result. The assertion we would clearly like to have is that $\Delta f = u$ on any open set Ω, for $u \in C^\infty(\Omega)$ and $f \in \mathcal{D}'(\Omega)$, implies $f \in C^\infty(\Omega)$. The hypotheses of 6.3.2 are restrictive in that the domain under consideration is all of $\mathbf{R}^n$, and in that u and f are required to satisfy integrability properties. The second question is to understand what the relevant properties of Δ are for this argument, and to see for how large a class of differential operators this type of argument will work. We will take up the second question first, and establish a regularity result of the type of 6.3.2 for a natural class of operators, and then return to the issue of obtaining a satisfactory local version for these operators (and in particular for Δ).

A cursory examination of the proof of 6.3.2 shows that it is basically an immediate consequence of 6.2.9(a) (given the Sobolev embedding theorem). Via the Fourier transform, the operator Δ is converted to multiplication by the polynomial $-|\xi|^2$. What we shall see is that operators whose associated polynomial shares some basic properties of the polynomial $|\xi|^2$ will have a property close enough to 6.2.9(a) to prove a regularity result akin to 6.3.2.

If $p : \mathbf{R}^n \to \mathbf{C}$ is a polynomial, we let $\mathrm{Re}(p)$ be the real part of p. Thus, $\mathrm{Re}(p) = \frac{1}{2}(p+\bar{p})$, so $\mathrm{Re}(p) : \mathbf{R}^n \to \mathbf{C}$ is a polynomial and if $p(\xi) = \sum c_\alpha \xi^\alpha$, then $\mathrm{Re}(p)(\xi) = \sum \mathrm{Re}(c_\alpha)\xi^\alpha$. If p is homogeneous of degree m, so is $\mathrm{Re}(p)$.

DEFINITION 6.3.3. *Suppose p is a polynomial on $\mathbf{R}^n$ which is homogeneous of degree m. Then p is called elliptic if $\xi \in \mathbf{R}^n$, $\xi \neq 0$, implies $p(\xi) \neq 0$. We say that p is strongly elliptic if $\mathrm{Re}(p)$ is elliptic. For $n \geq 2$, a strongly elliptic p will be called positive if $\mathrm{Re}(p)(\xi) > 0$ for all $\xi \neq 0$, and negative if $\mathrm{Re}(p)(\xi) < 0$ for all $\xi \neq 0$.*

Clearly any strongly elliptic polynomial is elliptic.

EXAMPLE 6.3.4: a) $p(\xi) = \sum_i^n \xi_i^2$ is strongly elliptic. More generally, $p(\xi) = \sum_{i=1}^n \xi_i^{2m}$ is strongly elliptic of order $2m$.
b) If p is a homogeneous real polynomial of degree 2 on $\mathbf{R}^2$, i.e., $p(X, Y) = aX^2 + bY^2 + cXY$, $(a, b, c \in \mathbf{R})$, then p is elliptic if and only if the non-empty level sets $p^{-1}(t)$, $t \neq 0$, are ellipses.
c) If $n \geq 2$ and p is real and elliptic of degree m, then m is even.

To see this, consider $p(\xi_1, 0, \ldots, 0)$. This is either identically 0, which is impossible by ellipticity, or equal to $c_1 \xi_1^m$ for some $c_1 \in \mathbf{R}$, $c_1 \neq 0$. Then $p(\xi_1, 1, 0, \ldots, 0)$ is a polynomial of degree m in ξ_1. If m is odd, this has a root, contradicting ellipticity.

d) On $\mathbf{R}^2$, let $p(X, Y) = X + iY$. Then p is of order 1 and is elliptic. However, it is not strongly elliptic.

DEFINITION 6.3.5. *Suppose*

$$D = \sum_{|\alpha| \leq m} a_\alpha(x) D^\alpha$$

is a differential operator of order m on an open set $\Omega \subset \mathbf{R}^n$. The symbol of D is the map $p : \Omega \to \{$ polynomials on $\mathbf{R}^n \}$, $x \mapsto p_x$, where p_x is the polynomial

$$p_x(\xi) = \sum_{|\alpha| = m} a_\alpha(x) \xi^\alpha.$$

(Thus, p_x is homogeneous of degree m for each $x \in \Omega$.) D is called a (strongly) elliptic differential operator of degree m if p_x is (strongly) elliptic of degree m for all $x \in \Omega$.

We remark that p_x depends only on the highest order terms of D, and hence so does ellipticity of D. We also remark that if Ω is connected, and D is strongly elliptic the symbol will be either positive for all $x \in \Omega$, or negative for all $x \in \Omega$.

EXAMPLE 6.3.6: a) Δ is strongly elliptic of degree 2. For any $m > 0$, Δ^m is strongly elliptic of degree $2m$. If p is any real polynomial of degree m, then $p(\Delta)$ is strongly elliptic of degree $2m$.

b) Let p_x be the symbol of D. Then the symbol of D^* is $(-1)^m \overline{p}_x$. Thus, if D is elliptic or strongly elliptic, D^* will have the same property.

c) If D_1, D_2 have symbols p_x and q_x respectively, then the symbol of $D_1 \circ D_2$ is $p_x(\xi) q_x(\xi)$. Thus if D_i are both elliptic, then $D_1 \circ D_2$ is also elliptic. If both are strongly elliptic, and one has real coefficients then $D_1 \circ D_2$ is strongly elliptic.

d) If D is elliptic, then D^*D is strongly elliptic. This follows from (b), (c).

e) If $D = \sum_{|\alpha| \leq m} c_\alpha D^\alpha$ is a differential operator with constant coefficients, then $p = p_x$ is constant in x. Under the Fourier transform, D corresponds to multiplication by a polynomial of the form $(-i)^m p + q$ where $\deg(q) < m$. More precisely, for $\varphi \in C_c^\infty(\mathbf{R}^n)$,

$$(D\varphi)^\wedge(\xi) = \left((-i)^m p(\xi) + q(\xi)\right)\widehat{\varphi}(\xi).$$

f) On $\mathbf{R}^2$, let $D = \partial_1 + i\partial_2$. Then by 6.3.4(d), D is elliptic but not strongly elliptic. We have $D^*D = -\Delta$.

Suppose now that p and q are elliptic polynomials of degree m. Then for $t \in \mathbf{R}$, $t \neq 0$, $p(t\xi)/q(t\xi) = p(\xi)/q(\xi)$. Thus, p/q can be viewed as a continuous function on the sphere, and in particular, it is bounded away from 0. That is, there is some $c \in \mathbf{R}$, $c > 0$, with $|p(\xi)| \geq c|q(\xi)|$ for all ξ. (The same is true of course if p, q are just continuous functions vanishing only at the origin and satisfying $f(tx) = t^m f(x)$ for $t > 0$.)

DEFINITION 6.3.7. *Suppose $\{p_x\}$ is a family of elliptic polynomials of degree m on $\mathbf{R}^n$, indexed by $x \in \Omega$, where Ω is any set. We say that $\{p_x\}$ is uniformly elliptic if for any fixed elliptic polynomial of degree m, say $q(\xi)$, there is a constant $c \in \mathbf{R}$, $c > 0$, such that $|p_x(\xi)| \geq c|q(\xi)|$ for all ξ and all $x \in \Omega$. We say that $\{p_x\}$ is uniformly strongly elliptic if $\{\mathrm{Re}(p_x)\}$ is uniformly elliptic.*

We remark that this is independent of the choice of q. Alternatively, uniform ellipticity of $\{p_x\}$ is equivalent to the existence of a positive constant c with $|p_x(\xi)| \geq c\sum_1^n |\xi_i|^m$.

DEFINITION 6.3.8. *If D is a differential operator of order m on Ω, we say that D is uniformly (strongly) elliptic on Ω if the image of the symbol map (i.e., $\{p_x \mid x \in \Omega\}$) is a uniformly (strongly) elliptic family of polynomials.*

EXAMPLE 6.3.9: a) If D is elliptic with constant (real) coefficients, then D is uniformly (strongly) elliptic.

b) If D is (strongly) elliptic on Ω and $V \subset \Omega$ is open with $\overline{V} \subset \Omega$.

and $\overline{V}$ compact, then D is uniformly (strongly) elliptic on V.
c) Let $D = \varphi(x)\Delta$ where $\varphi \in C^\infty(\mathbf{R}^n)$, $\varphi > 0$, and $\varphi(x) \to 0$ as $x \to \infty$. Then D is strongly elliptic but not uniformly elliptic.
d) If D is uniformly elliptic, so is D^*. This follows from 6.3.6(b).
e) If D_1, D_2 are uniformly elliptic, so is $D_1 \circ D_2$ (6.3.6(c)).

Example (b) above says roughly that elliptic operators are locally uniformly elliptic. Here is another version of this.

PROPOSITION 6.3.10. *Suppose D is a strongly elliptic operator of order m on $\Omega \subset \mathbf{R}^n$ (where Ω is open and connected). Suppose $V \subset \Omega$ and is open with $\overline{V} \subset \Omega$ and $\overline{V}$ compact. Then there is a uniformly strongly elliptic operator D_1 on $\mathbf{R}^n$ such that $D = D_1$ on V. Furthermore, all coefficients of D_1 can be chosen to lie in $BC^\infty(\mathbf{R}^n)$.*

PROOF: Let p_x be the symbol of D. We can assume $\text{Re}(p_x)$ is positive for all $x \in \Omega$. Choose $\psi \in C_c^\infty(\Omega)$, $0 \le \psi \le 1$, with $\psi = 1$ on $\overline{V}$. Let D_0 be any strongly elliptic operator of order m on $\mathbf{R}^n$ with constant coefficients. Then $D_1 = \psi D + (1 - \psi)D_0$ has symbol

$$r_x(\xi) = \psi(x)p_x(\xi) + (1 - \psi(x))q(\xi)$$

where q is the symbol of D_0. Thus, if $\text{Re}(q)$ is positive, which we can clearly arrange, D_1 has the required properties.

The fundamental property of elliptic operators that will enable us to generalize 6.3.2 is the following replacement for the properties of Δ given in 6.2.9.

THEOREM 6.3.11. (Garding's inequality) *Let D be a uniformly strongly elliptic operator of order $2m$ on $\mathbf{R}^n$ with all coefficients in $BC^\infty(\mathbf{R}^n)$. Then there are $a, b \in \mathbf{R}$ ($a \ge 0$ and $b > 0$ if $(-1)^m \text{Re}(p)$ is positive, where p is the symbol of D) such that for all φ in $C_c^\infty(\mathbf{R}^n)$,*

$$\text{Re}\big((aI + bD)\varphi, \varphi\big) \ge \|\varphi\|_m^2.$$

Of course 6.2.9(b) asserts that 6.3.11 holds for Δ with $a = 1, b = -1$. We shall see that Garding's inequality and the similar inequality for D^* will enable us to deduce an assertion similar

to 6.2.9(a) for D. This will then imply the regularity result for
D, generalizing 6.3.2. The proof of Garding's inequality will be
in three steps. First, we shall prove the case in which the coefficients are constant. This is quite easy and will follow by using
the Fourier transform and some simple inequalities for polynomials. The second step is to observe that locally the coefficients of
D are close to being constant. By comparing with the constant
coefficient case we shall then see that each $x \in \mathbf{R}^n$ has a neighborhood on which Garding's inequality holds for functions supported
in that neighborhood. The third step is to piece together the local
results by using a partition of unity. We shall now give the proof
in the constant coefficient case. As the other steps are a bit more
technically involved, we postpone that part of the argument until
an appendix. (Section 6.4).

PROOF OF 6.3.11 FOR CONSTANT COEFFICIENT D: Let p be the
symbol of D. Taking the Fourier transform and applying the
Plancherel theorem, we see that Garding's inequality is equivalent
to

$$\left(\left[a + b\left((-1)^m \operatorname{Re}(p)(\xi) + q(\xi) \right) \right] \widehat{\varphi}, \widehat{\varphi} \right) \geq \| (1 + |\xi|^2)^{m/2} \widehat{\varphi} \|^2$$

where $q(\xi)$ is a real polynomial of degree $\leq 2m - 1$. It therefore
suffices to find $a, b \in \mathbf{R}$ such that

$$a + b((-1)^m \operatorname{Re}(p)(\xi) + q(\xi)) \geq (1 + |\xi|^2)^m,$$

which is a simple exercise in polynomials. Namely, since $\operatorname{Re}(p)$
is elliptic of order $2m$ (and we can assume $(-1)^m \operatorname{Re}(p)$ is positive), it suffices (by the remarks preceding Definition 6.3.7) to take
$(-1)^m p(\xi) = |\xi|^{2m}$ (and to show we can take $b > 0$). But

$$\frac{|\xi|^{2m} + q(\xi)}{(1 + |\xi|^2)^m} = \frac{|\xi|^{2m} + q(\xi)}{|\xi|^{2m} + r(\xi)}$$

where q, r have degree at most $2m - 1$, and hence as $|\xi| \to \infty$, this
converges to 1. Thus for some R sufficiently large,

$$2\left(|\xi|^{2m} + q(\xi) \right) \geq (1 + |\xi|^2)^m \qquad \text{for } |\xi| \geq R.$$

Since both functions appearing in this inequality are continuous and $\{|\xi| \leq R\}$ is compact, it is clear that by choosing $a > 0$ sufficiently large we will have

$$a + 2(|\xi|^{2m} + q(\xi)) \geq (1 + |\xi|^2)^m \quad \text{for all } \xi.$$

This completes the proof.

We can now generalize 6.2.9(a).

THEOREM 6.3.12. *Let D be a uniformly strongly elliptic operator of order $2m$ on $\mathbf{R}^n$, with all coefficients in $BC^\infty(\mathbf{R}^n)$. Then for each $k \in \mathbf{Z}$, there is some $a_k \in \mathbf{R}$ such that*

$$a_k I + D : H^{k+2m}(\mathbf{R}^n) \to H^k(\mathbf{R}^n)$$

is an isomorphism of Hilbert spaces. In fact, if p is the symbol of D and $(-1)^m \operatorname{Re}(p)$ is positive, then we can choose a_k such that for all $c \geq a_k$

$$cI + D : H^{k+2m}(\mathbf{R}^n) \to H^k(\mathbf{R}^n)$$

is an isomorphism.

PROOF: We may clearly assume $(-1)^m \operatorname{Re}(p)$ is positive. We first claim it suffices to find constants c_k so that for all $c \geq c_k$, $cI + D : H^{k+2m}(\mathbf{R}^n) \to H^k(\mathbf{R}^n)$ is bounded below. For if this is the case we can find d_k such that $dI + D^* : H^{k+2m}(\mathbf{R}) \to H^k(\mathbf{R}^n)$ is bounded below for all $d \geq d_k$. (Recall that D^* is also uniformly strongly elliptic, with symbol $(-1)^{2m}\overline{p}$, so that its symbol also satisfies the positivity requirement.) Letting $a_k = \max\{c_k, d_{-k-2m}\}$, we have $c \geq a_k$ implies

$$cI + D : H^{k+2m}(\mathbf{R}^n) \to H^k(\mathbf{R}^n)$$

is bounded below, and so is its adjoint

$$cI + D^* : H^{-k}(\mathbf{R}^n) \to H^{-k-2m}(\mathbf{R}^n).$$

The theorem would then follow from Lemma 4.1.6.

To show the existence of such c_k we remark that Garding's inequality implies

$$\left|\big((cI + D)\varphi, \varphi\big)\right| \geq b^{-1}\|\varphi\|_m^2 \qquad \text{for any } c \geq a/b,$$

and hence (by definition of $\|\ \ \|_{-m}$) that

$$\|(cI + D)\varphi\|_{-m} \geq b^{-1}\|\varphi\|_m.$$

Thus, Garding's inequality immediately gives us $c_{-m} = a/b$ (cf. Proposition 6.2.5(c)). Suppose now that $k \geq 0$. To show the existence of c_{k-m}, we apply Garding's inequality to $(I - \Delta)^k D$, which is uniformly strongly elliptic of order $2(m + k)$, has coefficients in $BC^\infty(\mathbf{R}^n)$, and has symbol q with $(-1)^{k+m}q$ positive. Thus, we have positive constants d_1, d_2 such that

$$\mathrm{Re}((cI + (I - \Delta)^k D)\varphi, \varphi) \geq d_2\|\varphi\|_{k+m}^2 \qquad \text{for } c \geq d_1.$$

However,

$$
\begin{aligned}
((cI + (I - \Delta)^k D)\varphi, \varphi) &= c(\varphi, \varphi) + ((I - \Delta)^k D\varphi, \varphi) \\
&= c(\varphi, \varphi) + \langle D\varphi, \varphi\rangle_k \qquad \text{(by 6.2.9(b))}
\end{aligned}
$$

Thus, the real part is

$$
\begin{aligned}
&\leq \mathrm{Re}(c\langle\varphi, \varphi\rangle_k + \langle D\varphi, \varphi\rangle_k) \\
&\leq |\langle(cI + D)\varphi, \varphi\rangle_k| \\
&\leq \|(cI + D)\varphi\|_{k-m}\|\varphi\|_{k+m}.
\end{aligned}
$$

Therefore

$$d_2\|\varphi\|_{k+m}^2 \leq \|(cI + D)\varphi\|_{k-m}\|\varphi\|_{k+m},$$

and so

$$d_2\|\varphi\|_{k+m} \leq \|(cI + D)\varphi\|_{k-m}.$$

This shows the existence of c_{k-m}.

It remains to find c_{-k-m} for $k > 0$. As above, we can find $d_1, d_2 > 0$ such that $c \geq d_1$ implies

$$\mathrm{Re}\big((cI + D(I - \Delta)^k)\varphi, \varphi\big) \geq d_2\|\varphi\|_{k+m}^2.$$

Since $(I - \Delta)^k : H^{k+m}(\mathbf{R}^n) \to H^{-k+m}(\mathbf{R}^n)$ is an isometric isomorphism, for any $\psi \in C_c^\infty(\mathbf{R}^n)$ we can write $(I - \Delta)^k\varphi = \psi$ for a unique $\varphi \in H^{k+m}(\mathbf{R}^n)$. We remark that such a φ is automatically in $H^r(\mathbf{R}^n)$ for all r, and in particular, it is smooth. (Namely, since $\psi \in H^s(\mathbf{R}^n)$ for all s, for any r we can find $\theta \in H^r(\mathbf{R}^n)$ with $(1 - \Delta)^k\theta = \psi$, by Proposition 6.2.9(a). Since this is also an equation in $H^{-k+m}(\mathbf{R}^n)$ if $r \geq k + m$, we have $\theta = \varphi$.) By a continuity argument, the above consequence of Garding's inequality for $D(I - \Delta)^k$ also holds for $\varphi \in \bigcap_{r \in \mathbf{Z}} H^r(\mathbf{R}^n)$. We then have

$$\begin{aligned}
d_2\|\psi\|_{-k+m}^2 &\leq \mathrm{Re}(c(\varphi, \varphi) + (D\psi, \varphi)) \\
&\leq \mathrm{Re}(c\langle\varphi, \varphi\rangle_k + (D\psi, \varphi)) \\
&= \mathrm{Re}(c((1 - \Delta)^k\varphi, \varphi) + (D\psi, \varphi)) \\
&= \mathrm{Re}((cI + D)\psi, \varphi) \\
&\leq |((cI + D)\psi, \varphi)| \\
&\leq \|(cI + D)\psi\|_{-k-m}\|\varphi\|_{k+m} \\
&= \|(cI + D)\psi\|_{-k-m}\|\psi\|_{-k+m}
\end{aligned}$$

Hence

$$d_2\|\psi\|_{-k+m} \leq \|(cI + D)\psi\|_{-k-m}$$

as required. This completes the proof of 6.3.12.

As a consequence we deduce our sought-after generalization of 6.3.2.

THEOREM 6.3.13. *Let D be a uniformly strongly elliptic operator of order $2m$ on $\mathbf{R}^n$ with all coefficients in $BC^\infty(\mathbf{R}^n)$. Suppose $Df = u$ where $u \in H^r(\mathbf{R}^n)$ for all r (and in particular is smooth). If $f \in H^k(\mathbf{R}^n)$ for some $k \in \mathbf{Z}$ (e.g., $f \in L^2(\mathbf{R}^n)$),then $f \in C^\infty(\mathbf{R}^n)$.*

PROOF: We argue exactly as in the proof of 6.3.2. Let a_k be as in in 6.3.12. Fix $c > a_k$, a_{k-2m}. Then $(cI+D)f = cf+u$ in $H^{k-2m}(\mathbf{R}^n)$. Since $cf + u \in H^k(\mathbf{R}^n)$, by 6.3.12 there is a unique $g \in H^{k+2m}(\mathbf{R}^n)$ such that $(cI + D)g = cf + u$. However, this is also an equation in $H^{k-2m}(\mathbf{R}^n)$, so $f = g$. It follows that $f \in H^{k+2m}(\mathbf{R}^n)$. Arguing inductively, we have $f \in H^r(\mathbf{R}^n)$ for all r, and hence by the Sobolev embedding theorem (5.2.4), $f \in C^\infty(\mathbf{R}^n)$.

We now turn to the problem of giving a local version of Theorem 6.3.13. At the same time, we can relax the hypotheses on D. Here is the basic elliptic regularity theorem.

THEOREM 6.3.14. (Elliptic Regularity) *Let D be an elliptic operator of order m on an open set $\Omega \subset \mathbf{R}^n$. Suppose $T \in \mathcal{D}'(\Omega)$ and $DT = u$ where $u \in C^\infty(\Omega)$. Then $T \in C^\infty(\Omega)$.*

The proof will be very similar to that of 6.3.13. The basic additional ingredient is, not surprisingly, to consider $D(\psi T)$ where $\psi \in C_c^\infty(\Omega)$. If we had $D(\psi T) = \psi u$ (which of course we will not in general) we would be in a position to apply 6.3.13 directly. However, while $D(\psi T) - \psi u$ is not 0, it is a lower order differential operator applied to T. We shall see that this lower order operator does not essentially disturb the proof of 6.3.13. We now turn to the details. We begin with a simple reduction.

LEMMA 6.3.15. *To prove 6.3.14 it suffices to assume $T \in H_{\mathrm{loc}}^k(\Omega)$ for some k.*

PROOF: Write $\Omega = \cup V_i$ where $V_i \subset \overline{V}_i \subset \Omega$ with V_i open and $\overline{V}_i$ compact. It clearly suffices to see (cf. exercise 6.2) that $T|V_i$ is a C^∞-function for each i. Let $V = V_i$. Choose $\psi \in C_c^\infty(\Omega)$ with $\psi = 1$ on V. Then $\psi T \in \mathcal{D}_c'(\Omega)$ and hence $\psi T \in H^k(\Omega)$ for some k (Proposition 6.2.3(a).) In particular $(\psi T)|V \in H_{\mathrm{loc}}^k(V)$. Since $\psi = 1$ on V, $(\psi T)|V = T|V$, so $T|V \in H_{\mathrm{loc}}^k(V)$. But on V we still have $D(T|V) = u$, so if we have established 6.3.14 for distributions in H_{loc}^k, we have $T|V$ is C^∞, verifying the lemma.

PROOF OF 6.3.14: We first observe that D^*D is strongly elliptic (by 6.3.6(d)) and of order $2m$. We have $(D^*D)T = D^*u \in C^\infty(\Omega)$. Therefore, it suffices to prove the theorem for D^*D. Hence, we may assume D is strongly elliptic and of order $2m$. We may also assume $(-1)^m \operatorname{Re}(p)$ is positive, where p is the symbol of D.

By the lemma we may assume $T \in H^k_{\text{loc}}(\Omega)$. We shall show that $T \in H^r_{\text{loc}}(\Omega)$ for all r, and hence by the Sobolev theorem (6.2.12) this will show T is smooth. Let $\psi \in C^\infty_c(\Omega)$. We then have $\psi D(T) = \psi u$. Writing M_ψ for multiplication by ψ we may write the differential operator $M_\psi \circ D = D \circ M_\psi - D_1$ where D_1 is of order strictly less than $2m$, and all of whose coefficients have support contained in $\operatorname{supp}(\psi)$. (Cf. exercise 1.26.) That is, $\psi D(T) = D(\psi T) - D_1(T)$, so $D(\psi T) = \psi u + D_1(T)$. We remark that $D_1(T) \in \mathcal{D}'_c(\Omega)$. Let $\widetilde{D}$ be a uniformly strongly ellipitic operator on $\mathbf{R}^n$ with all coefficients in $BC^\infty(\mathbf{R}^n)$ such that $\widetilde{D} = D$ on a neighborhood of $\operatorname{supp}(\psi)$ (Proposition 6.3.10). Then $\widetilde{D}(\psi T) = \psi u + D_1(T)$. We now argue as in the proof of 6.3.13. Since $T \in H^k_{\text{loc}}(\Omega)$, this is an equation in $H^{k-2m}(\mathbf{R}^n)$ (cf. Proposition 6.2.11). Let a_k be the constants for $\widetilde{D}$ given by Theorem 6.3.12. Let $c \geq a_{k-2m}, a_{k-2m+1}$. Then

$$(cI + \widetilde{D})(\psi T) = \psi u + D_1(T) + c\psi T.$$

The right side of the equation lies in $H^{k-2m+1}(\mathbf{R}^n)$ since D_1 is an operator of order at most $2m-1$. (We are implicitly using 6.2.3(c) and 6.2.11(a),(b).) By choice of c there is a unique $g \in H^{k+1}(\mathbf{R}^n)$ such that

$$(cI + \widetilde{D})g = \psi u + D_1(T) + c\psi T.$$

This is also an equation in H^{k-2m} and hence $\psi T = g$. Thus, $\psi T \in H^{k+1}(\mathbf{R}^n)$, i.e., $T \in H^{k+1}_{\text{loc}}(\Omega)$. Continuing inductively, this shows $T \in H^r_{\text{loc}}(\Omega)$ for all r, and this completes the proof.

Here is an immediate consequence of Theorem 6.3.14.

COROLLARY 6.3.16. *If D is an elliptic operator on Ω, and T is an eigendistribution of D, (i.e., $DT = \lambda T$ for some $\lambda \in \mathbf{C}$), then $T \in C^\infty(\Omega)$.*

PROOF: $(D - \lambda I)T = 0$, and $D - \lambda I$ is elliptic.

6.4. Appendix to 6.3: proof of Garding's inequality

We have proven Garding's inequality (6.3.11) for constant co-efficient operators and now present the proof for the general case. The general idea of the proof is discussed following the statement of 6.3.11. To apply the constant coefficient case in the proof of the general case we need to examine the proof in the constant co-efficient case to understand the dependence of the constants a, b appearing in the statement of Garding's inequality on the properties of D. An examination of the proof for constant coefficients easily shows:

LEMMA 6.4.1. *Suppose* $\{D_s \mid s \in S\}$ *is a family of constant co-efficient strongly elliptic operators of order* $2m$ *on* $\mathbf{R}^n$, *indexed by some set* S. *Suppose*

   i) $\{D_s\}$ *is a uniformly strongly elliptic family (i.e.,* $\{\mathrm{Re}(p_s)\}$ *is uniformly elliptic where* p_s *is the symbol of* D_s*).*
  ii) *The coefficients of* D_s *are uniformly bounded over* S.
 iii) $(-1)^m \mathrm{Re}(p_s)$ *is positive for all* s.
      Then there are $a, b > 0$ *such that*

$$\mathrm{Re}\big((aI + bD_s)\varphi, \varphi\big) \geq \|\varphi\|_m^2 \quad \text{for all } \varphi \in C_c^\infty(\mathbf{R}^n), \text{ and all } s \in S$$

We now present 4 lemmas each of which presents an inequality of a general nature, that is, concerning Sobolev spaces or differential operators, not specifically elliptic operators.

LEMMA 6.4.2. *For any* $m \geq 1$ *and for any* $\varepsilon > 0$, *there is a constant* $c(\varepsilon) \in \mathbf{R}$ *such that*

$$\|\varphi\|_{m-1}^2 \leq \varepsilon \|\varphi\|_m^2 + c(\varepsilon)\|\varphi\|_0^2 \quad \text{for all } \varphi \in C_c^\infty(\mathbf{R}^n).$$

PROOF: $\|\varphi\|_m^2 = ((1 + |\xi|^2)^m \widehat{\varphi}, \widehat{\varphi})$, so it suffices to find $c(\varepsilon)$ such that

$$(1 + |\xi|^2)^{m-1} \leq \varepsilon(1 + |\xi|^2)^m + c(\varepsilon).$$

But for R sufficiently large, $|\xi| \geq R$ implies

$$(1 + |\xi|^2)^{m-1} \leq \varepsilon(1 + |\xi|^2)^m,$$

and therefore we can take $c(\varepsilon) = (1 + R^2)^{m-1}$.

In the statement of Garding's inequality we are concerned with an expression of the form $(D\varphi, \varphi)$ where D is a differential operator. The next three lemmas are concerned with some general facts about this expression.

Suppose $D = a(x)D^\alpha$. If we write $\alpha = \beta + \gamma$, then we have

$$a(x)D^\alpha = a(x)D^\beta D^\gamma = D^\beta(a(x)D^\gamma) + \widetilde{D}D^\gamma$$

where $\text{order}(\widetilde{D}) < |\beta|$, and the coefficients of $\widetilde{D}$ depend on the derivatives of $a(x)$ up to order $|\beta|$. It follows that we can write

$$(a(x)D^\alpha\varphi, \psi) = \left(D^\beta(a(x)D^\gamma)\varphi, \psi\right) + (\widetilde{D}D^\gamma\varphi, \psi)$$
$$= (-1)^{|\beta|}\left(a(x)D^\gamma\varphi, D^\beta\psi\right) + (D'\varphi, \psi)$$

where $\text{order}(D') < |\alpha|$ and the coefficients of D' depend on the derivatives of $a(x)$ up to order $|\beta|$. In fact, if $c_\lambda(x)D^\lambda$ appears in D', then $c_\lambda(x)$ depends only on the derivatives of $a(x)$ up to order $|\alpha| - |\lambda|$. Thus

$$|(a(x)D^\alpha\varphi, \psi)| \leq c\|\varphi\|_{|\alpha|}\|\psi\|_{|\beta|} + |(D'\varphi, \psi)|.$$

Expressing any differential operator D as a sum of terms of the form $a_\alpha(x)D^\alpha$ and arguing by induction on the order of D, we obtain:

LEMMA 6.4.3. *Suppose D is a differential operator of order m on an open $\Omega \subset \mathbf{R}^n$, and suppose the coefficients are all in $BC^m(\Omega)$. Write $m = k + \ell$ where $k, \ell \geq 0$. Then there is a constant c such that for all $\varphi, \psi \in C_c^\infty(\Omega)$,*

$$|(D\varphi, \psi)| \leq c\|\varphi\|_k\|\psi\|_\ell$$

where c depends only on the norms of the coefficients of D in $BC^m(\Omega)$. (In particular, c does not otherwise depend on Ω.)

We will need the following two special cases.

LEMMA 6.4.4. *Fix n and m. If $\varepsilon > 0$, then there is a $\delta > 0$ so that for every differential operator D of order $2m$ on an open set $\Omega \subset \mathbf{R}^n$ with all coefficients of D having $BC^{2m}(\Omega)$-norm at most δ, we have*

$$|(D\varphi, \varphi)| \leq \varepsilon \|\varphi\|_m^2 \quad \text{for all } \varphi \in C_c^\infty(\Omega).$$

The proof of Lemma 6.4.4 is immediate from Lemma 6.4.3.

LEMMA 6.4.5. *If D is a differential operator of order $2m - 1$ on $\Omega \subset \mathbf{R}^n$ with coefficients in $BC^\infty(\Omega)$, then for any $\varepsilon > 0$ there is a constant $B(\varepsilon)$ such that for all $\varphi \in C_c^\infty(\Omega)$ we have*

$$|(D\varphi, \varphi)| \leq \varepsilon \|\varphi\|_m^2 + B(\varepsilon)\|\varphi\|_0^2.$$

The constant $B(\varepsilon)$ depends only on the norms of the coefficients of D in $BC^{2m-1}(\Omega)$.

PROOF: By Lemma 6.4.3 there is $c \in \mathbf{R}$ depending only on the norms of the coefficients of D in $BC^{2m-1}(\Omega)$ such that

$$|(D\varphi, \varphi)| \leq c\|\varphi\|_m\|\varphi\|_{m-1}$$

Since for any $r > 0$, and any $a, b \geq 0$ we have $ab \leq \frac{1}{2}(ra^2 + \frac{1}{r}b^2)$, it follows that for any $\varepsilon > 0$ there is $A(\varepsilon) > 0$ depending only on (ε and) the norms of the coefficients of D in $BC^{2m-1}(\Omega)$ such that for all $\varphi \in C_c^\infty(\Omega)$ we have

$$|(D\varphi, \varphi)| \leq \frac{\varepsilon}{2}\|\varphi\|_m^2 + A(\varepsilon)\|\varphi\|_{m-1}^2.$$

An application of Lemma 6.4.2 then proves 6.4.5.

With these general estimates in hand we now turn to the proof of Garding's inequality.

PROOF OF 6.3.11: The next step is to use the result for constant coefficients to obtain a suitable local version. For each $s \in$

$\mathbf{R}^n$, let D_s be the differential operator with constant coefficients $c_\alpha(s)$ where $D = \sum_{|\alpha| \le 2m} c_\alpha(s) D^\alpha$. Since $\{D_s \mid s \in \mathbf{R}^n\}$ satisfies the hypotheses of Lemma 6.4.1, we can find $a, b \ge 1$ such that $\operatorname{Re}((aI + bD_s)\varphi, \varphi) \ge \|\varphi\|_m^2$ for all $s \in \mathbf{R}^n$. Now choose δ so that Lemma 6.4.4 holds with $\varepsilon = 1/2b$. Since $c_\alpha \in BC^\infty(\mathbf{R}^n)$, all the derivatives of all c_α are uniformly continuous. It follows that there is $r > 0$ such that for any $s \in \mathbf{R}^n$ the differential operator $D - D_s$ has coefficients in $BC^{2m}(\{|x_i - s_i| < r \text{ for all } i\})$ of norm at most δ. Let U_s be the open cube of side r centered at s. We then have:

LEMMA 6.4.6. *There are $c, d > 0$ such that for any $s \in \mathbf{R}^n$ and any $\varphi \in C_c^\infty(U_s)$ we have*

$$\operatorname{Re}((cI + dD)\varphi, \varphi) \ge \|\varphi\|_m^2.$$

PROOF:

$$\left|((aI + bD)\varphi, \varphi) - (aI + bD_s)\varphi, \varphi))\right| = b|((D - D_s)\varphi, \varphi)|$$
$$\le \frac{1}{2}\|\varphi\|_m^2 \quad \text{by Lemma 6.4.4 .}$$

Hence,

$$\operatorname{Re}((aI + bD_s)\varphi, \varphi) \ge \frac{1}{2}\|\varphi\|_m^2,$$

so we can take $c = 2a, d = 2b$.

We now turn to the final step in the proof of 6.3.11, namely piecing together the local result for the various U_s via a suitable partition of unity. We now construct such a partition of unity with certain uniform properties.

Let V_s be the open cube of side $\frac{3}{4}r$ centered at s. Let $S = \frac{r}{2}\mathbf{Z}^n \subset \mathbf{R}^n$. Thus $\{U_s\}_{s \in S}$ and $\{V_s\}_s \in S$ are locally finite covers of $\mathbf{R}^n$. Fix $\lambda_0 \in C_c^\infty(U_0)$ with $0 \le \lambda_0 \le 1$ and $\lambda_0 = 1$ on V_0. For each $s \in S$, let λ_s be the translation of λ_0 by s; i.e., $\lambda_s(x) = \lambda_0(x - s)$. Then $\psi = \sum_{s \in S} \lambda_s$ is a well-defined C^∞-function and $\psi(x) > 0$ for all $x \in \mathbf{R}^n$. Let $\psi_s = (\lambda_s/\psi)^{1/2}$. Then $\psi_s \in C_c^\infty(U_s)$, $0 \le \psi_s \le 1$, and $\sum \psi_s(x)^2 = 1$ for all $x \in \mathbf{R}^n$. Thus, $\{\psi_s^2\}_{s \in S}$ is a partition of unity subordinate to $\{U_s\}$, but it also has the property that for each q, $\{\psi_s\}_{s \in S}$ is uniformly bounded in the $BC^q(\mathbf{R}^n)$-norm. This is immediate from the observation that $\psi_s(x) = \psi_0(x - s)$.

Now choose c, d as in Lemma 6.4.6. Let $\varphi \in C_c^\infty(\mathbf{R}^n)$. We may then apply 6.4.6 to each $\psi_s\varphi$. Summing over s, we obtain

$$(1) \qquad \sum_{s \in S} \mathrm{Re}\big((cI + dD)\psi_s\varphi, \psi_s\varphi\big) \geq \sum_s \|\psi_s\varphi\|_m^2.$$

We shall now compare the left side of this equation to $\mathrm{Re}\big((cI + dD)\varphi, \varphi\big)$ and the right side to $\|\varphi\|_m^2$. This will then yield the required inequality. Both of these comparisons will use the same technique, namely, an application of Lemma 6.4.5.

We claim first that there is a constant K_1 such that for all $\varphi \in C_c^\infty(\mathbf{R}^n)$,

$$(2) \qquad \sum_s \|\psi_s\varphi\|_m^2 \geq \frac{3}{4}\|\varphi\|_m^2 - K_1\|\varphi\|_0^2.$$

Namely, we have

$$\|\varphi\|_m^2 = ((I - \Delta)^m\varphi, \varphi) = \Big(\sum_{s \in S} \psi_s^2(I - \Delta)^m\varphi, \varphi\Big).$$

On the other hand

$$\sum_{s \in S} \|\psi_s\varphi\|_m^2 = \sum_{s \in S}\big((I-\Delta)^m(\psi_s\varphi), \psi_s\varphi\big) = \sum_{s \in S}\big(\psi_s(I-\Delta)^m(\psi_s\varphi), \varphi\big)$$

Therefore

$$(3) \qquad \|\varphi\|_m^2 - \sum_{s \in S}\|\psi_s\varphi\|_m^2 = (D_1\varphi, \varphi)$$

where D_1 is the differential operator

$$D_1 = \sum_{s \in S}\big\{M_{\psi_s^2}(I - \Delta)^m - M_{\psi_s}(I - \Delta)^m M_{\psi_s}\big\}.$$

Then D_1 is defined on all $\mathbf{R}^n$, all coefficients are in $BC^\infty(\mathbf{R}^n)$, and order$(D_1) \leq 2m-1$. Therefore by Lemma 6.4.5 there is a constant $K_1 > 0$ such that for all $\varphi \in C_c^\infty(\mathbf{R}^n)$,

$$(4) \qquad |(D_1\varphi, \varphi)| \leq \frac{1}{4}\|\varphi\|_m^2 + K_1\|\varphi\|_0^2.$$

Combining equations (3) and (4), we obtain (2).

We now make a similar argument to show that we can find $K_2 > 0$ such that for all $\varphi \in C_c^\infty(\mathbf{R}^n)$,

$$(5)$$

$$\left\{ \sum_{s \in S} \mathrm{Re}\big((cI + dD)(\psi_s\varphi), \psi_s\varphi\big) \right\} - \mathrm{Re}\big((cI + dD)\varphi, \varphi\big)$$

$$\leq \frac{1}{4}\|\varphi\|_m^2 + K_2\|\varphi\|_0^2.$$

Namely, the left side is

$$\mathrm{Re}\Big(\sum_{s \in S} \psi_s(cI + dD)(\psi_s\varphi), \varphi\Big) - \mathrm{Re}\Big(\sum_{s \in S} \psi_s^2(cI + dD)\varphi, \varphi\Big)$$

$$= \mathrm{Re}(D_2\varphi, \varphi)$$

where D_2 is the differential operator

$$D_2 = \sum_{s \in S} \left\{ M_{\psi_s}(cI + dD)M_{\psi_s} - M_{\psi_s^2}(cI + dD) \right\}.$$

As above, this operator is defined on all $\mathbf{R}^n$, has $BC^\infty(\mathbf{R}^n)$ coefficients, and order$(D_2) \leq 2m - 1$. Thus, equation (5) follows immediately from Lemma 6.4.5.

Combining equations (1), (2), (5) yields

$$\mathrm{Re}((cI + dD)\varphi, \varphi) + \frac{1}{4}\|\varphi\|_m^2 + K_2\|\varphi\|_0^2 \geq \frac{3}{4}\|\varphi\|_m^2 - K_1\|\varphi\|_0^2.$$

Thus,

$$\mathrm{Re}\big((c'I + dD)\varphi, \varphi\big) \geq \frac{1}{2}\|\varphi\|_m^2$$

where $c' = c + K_1 + K_2$. Thus, for all $\varphi \in C_c^\infty(\mathbf{R}^n)$,

$$\mathrm{Re}\big((2c'I + 2dD)\varphi, \varphi\big) \geq \|\varphi\|_m^2,$$

and this proves Garding's inequality in general.

6.5. A spectral theorem for elliptic operators

In this section we apply the spectral theorem for compact operators (Theorem 3.2.3) to deduce a spectral theorem for elliptic differential operators. In addition to 3.2.3, we make use of Rellich's theorem (5.2.8), Garding's inequality (6.3.11), and elliptic regularity (6.3.14).

We recall that an orthonormal basis of $L^2([-\pi, \pi])$ (with normalized Lebesgue measure) is given by $e_n(x) = e^{inx}$, $n \in \mathbf{Z}$. In section 3.3 we discussed this as a natural choice of basis for $L^2([-\pi, \pi])$, as they are simultaneous eigenvectors for the translation operators on the circle group under the natural identification $L^2([-\pi, \pi]) \cong L^2(S^1)$, and saw how to generalize this to other compact groups. Here we observe that we also have that e_n are (C^∞) eigenvectors of the differential operator $D = \partial^2/\partial x^2$ on $(-\pi, \pi)$. D is of course just the one-dimensional Laplace operator, and is elliptic. This section is devoted to proving the following result which implies the presence of an orthonormal basis of eigenvectors in a much more general setting.

THEOREM 6.5.1. *Suppose $\Omega \subset \mathbf{R}^n$ is open and bounded. Let D be a formally self-adjoint strongly elliptic differential operator of order $2m$ defined on a neighborhood of $\overline{\Omega}$. Then there is a sequence $\varphi_j \in C^\infty(\Omega) \cap H^m(\Omega)$ such that*

i) *$\{\varphi_j\}$ is an orthogonal basis of $L^2(\Omega)$; with a suitable choice of inner product on $H^m(\Omega)$, $\{\varphi_j\}$ is an orthonormal basis of $H^m(\Omega)$.*

ii) *There are $\lambda_j \in \mathbf{R}$ with $D\varphi_j = \lambda_j \varphi_j$.*

iii) *$|\lambda_j| \to \infty$.*

iv) $\{\lambda_j\}$ exhausts the set of eigenvalues for D with eigenfunctions in $H^m(\Omega)$; for each such eigenvalue, the eigenspace in $H^m(\Omega)$ is finite dimensional.

REMARK: There may be many other eigenvectors of D in $L^2(\Omega)$ with eigenvalue not in $\{\lambda_j\}$, but where the eigenvectors do not lie in $H^m(\Omega)$. A simple example appears in 6.5.2 below. This is a reflection of the following fact (which we do not prove): If Ω has a sufficiently smooth boundary and φ is a function which is smooth on a neighborhood of $\overline{\Omega}$, then $\varphi \in H^m(\Omega)$ if and only if $D^\alpha \varphi = 0$ on the boundary of Ω, for all $|\alpha| < m$. One can, in fact, develop a theory of boundary values for functions in $L^{2,m}(\Omega)$, in which case $H^m(\Omega)$ consists exactly of those functions with all derivatives below order m vanishing of the boundary. Thus, Theorem 6.5.1 says not only can we find an orthonormal basis of $L^2(\Omega)$ consisting of eigenvectors of D, but that these eigenfunctions can be chosen to lie in a space of functions with some given boundary behavior. This raises the question as to when one can choose $\{\varphi_j\}$ to be an orthonormal basis of $L^2(\Omega)$, eigenfuctions of D, and satisfying some other type of boundary condition. This is one question that arises in the study of boundary value problems for partial differential equations.

EXAMPLE 6.5.2: i) Let $\Omega = (0, T) \subset \mathbf{R}$, and $D = \partial^2/\partial t^2$. Then we can take $\varphi_n(t) = \sin(n\pi t/T)$, $n > 0$. The eigenvalues are $\lambda_n = -n^2\pi^2/T^2$. For any $\lambda \in \mathbf{R}$, $\varphi(t) = \sin(\lambda t)$ will be an eigenfunction of D in $L^2(\Omega)$ with eigenvalue $-\lambda^2$. Furthermore, for any $a, b \in \mathbf{R}$, $\psi(t) = at + b$ will be in the kernel of D, i.e., is an eigenfunction with eigenvalue 0. However, as one may verify (exercise 6.13), none of these other eigenfunctions lie in $H^1(\Omega)$.
ii) Let $D = \Delta$ on $\mathbf{R}^2$. If $\Omega = (0, 1) \times (0, 1)$, then the functions $\varphi_{n,k}(s, t) = \sin(\pi n s)\sin(\pi k t)$, $n, k > 0$, are an orthonormal basis of $L^2(\Omega)$ consisting of eigenfunctions lying in $H^1(\Omega)$. The eigenvalues are $-\pi(n^2 + k^2)$. For a general Ω, it is not easy to compute explicitly the eigenvectors and eigenvalues.

We begin the proof of Theorem 6.5.1 with the following direct consequence of Garding's inequality. For $D = \Delta$, these facts appear in Proposition 6.2.5.

PROPOSITION 6.5.3. *Let D and Ω satisfy the hypotheses of Theorem 6.5.1. Choose (using 6.3.10) $a, b \in \mathbf{R}$ such that Garding's inequality (6.3.11) holds for all $\varphi \in C_c^\infty(\Omega)$. Then:*

i) *$aI + bD : H^m(\Omega) \to H^{-m}(\Omega)$ is an isomorphism.*

ii) *Define $B : H^m(\Omega) \times H^m(\Omega) \to \mathbf{C}$ by $B(\varphi, \psi) = ((aI + bD)\varphi, \psi)$. Then B is an inner product on $H^m(\Omega)$ that defines a norm equivalent to the standard one.*

PROOF: i) Since D is of order $2m$, the map in question is bounded. For $\varphi \in C_c^\infty(\Omega)$ we have by definition that

$$\|(aI + bD)\varphi\|_{-m} = \sup\{|((aI + bD)\varphi, \psi)| \ \big| \ \|\psi\|_m = 1\}.$$

Letting $\psi = \varphi/\|\varphi\|_m$ and applying Garding's inequality (6.3.11), we obtain $\|(aI + bD)\varphi\|_{-m} \geq \|\varphi\|_m$. Thus $aI + bD$ is bounded below. To see $aI + bD$ is an isomorphism, by 4.1.6 it suffices to see that $(aI + bD)^* : H^{-m}(\Omega)' \to H^m(\Omega)'$ is also bounded below. However, since D is formally self-adjoint, this map can be identified with $aI + bD : H^m(\Omega) \to H^{-m}(\Omega)$, which we have just seen is bounded below.

ii) B is sesquilinear since $D = D^*$. Since order$(D) = 2m$, we can write $aI + bD = \sum_i D_{i,1} \circ D_{i,2}$ where all $D_{i,1}$ and $D_{i,2}$ are of order at most m. Then if $\varphi, \psi \in C_c^\infty(\Omega)$, $B(\varphi, \psi) = \sum_i (D_{i,2}\varphi, D_{i,1}^*\psi)$, so $|B(\varphi, \psi)| \leq c\|\varphi\|_m \|\psi\|_m$ for some c. In particular, $B(\varphi, \varphi) \leq c\|\varphi\|_m^2$. On the other hand, Garding's inequality implies $B(\varphi, \varphi) \geq \|\varphi\|_m^2$, verifying the assertion.

PROOF OF THEOREM 6.5.1: By 6.5.3, we can define $A : H^{-m}(\Omega) \to H^m(\Omega)$ by $A = (aI + bD)^{-1}$. Let $i : H^m(\Omega) \to H^{-m}(\Omega)$ be the inclusion. Then $T = A \circ i : H^m(\Omega) \to H^m(\Omega)$ is a bounded operator. Since Ω is bounded, Rellich's theorem (5.2.8) (and Proposition 3.1.14) implies that i, and hence T, is compact. We claim that T is self-adjoint with respect to the inner product B of 6.5.2. Namely, for $\varphi, \psi \in C_c^\infty(\Omega)$,

$$B(T\varphi, \psi) = \Big((aI + bD)(A \circ i)\varphi, \psi\Big) = (\varphi, \psi).$$

Hence we also have

$$B(\varphi, T\psi) = \overline{B(T\psi, \varphi)} = \overline{(\psi, \varphi)} = (\varphi, \psi).$$

Therefore T is self-adjoint and compact. We now apply the spectral theorem (3.2.3) to obtain an orthonormal basis $\{\varphi_j\}$ of $H^m(\Omega)$ such that $T\varphi_j = \beta_j\varphi_j$, where $\beta_j \in \mathbf{R}$ and $\beta_j \to 0$ as $j \to \infty$. Since i and A are both injective, so is T, and therefore for each j we have $\beta_j \neq 0$. Applying $aI + bD$ to the equation $T\varphi_j = \beta_j\varphi_j$, we deduce $\varphi_j = \beta_j(aI + bD)\varphi_j$, and thus $D\varphi_j = \lambda_j\varphi_j$ where $\lambda_j = \frac{1-a\beta_j}{b\beta_j}$. Since $\beta_j \to 0$, $|\lambda_j| \to \infty$. Since D is elliptic, Corollary 6.3.16 implies $\varphi_j \in C^\infty(\Omega)$.

We know $\{\varphi_j\}$ is an orthonormal basis of $H^m(\Omega)$. We claim it is also an orthogonal basis of $L^2(\Omega)$, i.e., $\varphi_j/\|\varphi_j\|_0$ is an orthonormal basis of $L^2(\Omega)$. Since the finite linear combination of $\{\varphi_j\}$ are dense in $H^m(\Omega)$, they are also dense in $L^2(\Omega)$, and hence (Proposition A.26) it suffices to see $\{\varphi_j\}$ are mutually orthogonal in $L^2(\Omega)$. For $j \neq k$ we have

$$0 = B(\varphi_j, \varphi_k) = \Big((aI + bD)\varphi_j, \varphi_k\Big) = \beta_j^{-1}(\varphi_j, \varphi_k).$$

Thus, $\varphi_j \perp \varphi_k$ in L^2.

To complete the proof of Theorem 6.5.1, it only remains to prove assertion (iv). Suppose $D\varphi = \lambda\varphi$ where $\varphi \in H^m(\Omega)$, $\varphi \neq 0$. Then $(aI + bD)\varphi = (a + b\lambda)\varphi$, and since $aI + bD$ is injective on $H^m(\Omega)$ (6.5.3(a)), $a + b\lambda \neq 0$. Thus, $(a + b\lambda)^{-1}$ is an eigenvalue of T, so $(a + b\lambda)^{-1} = \beta_j$ for some j, and therefore $\lambda = \lambda_j$. Furthermore, φ is an eigenvector of T with eigenvalue β_j, and this space of eigenvectors is finite dimensional by 3.2.3. This completes the proof.

REMARK 6.5.4: We conclude this section with a few incomplete but hopefully suggestive comments regarding the extension of this result to compact manifolds.

1) If M is a compact smooth manifold we can define Sobolev spaces and differential operators on M. If M has no boundary, then $L^{2,m}(M) = H^m(M)$, and the discussion of boundary values in the remark following the statement of 6.5.1 is not relevant. One can then deduce that if D is a formally self-adjoint strongly elliptic operator on a compact smooth manifold (with a given volume form), then there is an orthonormal basis $\{\varphi_j\}$ of $L^2(M)$ consisting of eigenvectors of D, each φ_j is smooth, $|\lambda_j| \to \infty$, and $\{\lambda_j\}$

exhausts the eigenvalues of D for any eigendistribution. In particular, every eigenspace is finite dimensional. If M is the circle and $D = d^2/d\theta^2$, the eigenvectors are the standard orthonormal basis $\varphi_n(\theta) = e^{in\theta}$.

2) If M is a compact Riemannian manifold, then the Riemannian metric defines in a canonical way a second order formally self-adjoint strongly elliptic operator Δ on M called the Laplace-Beltrami operator on M. The eigenvalues of Δ (which form a discrete subset of $\mathbf{R}$) are thus geometric invariants of M, and there are a number of interesting relations between these and other geometric properties of M. In a similar way, if $\Omega \subset \mathbf{R}^n$ is open and bounded, the eigenvalues of Δ appearing in Theorem 6.5.1 are naturally attached to Ω, and are a reflection of the "shape" of Ω.

3) In certain cases the Peter-Weyl theorem (3.3.4) and the spectral theorem for elliptic operators are closely connected. Namely, suppose G is a compact Lie group. Then there is a G-invariant Riemannian metric on G. If we let Δ be the associated Laplace-Beltrami operator on G, then every $g \in G$ acting by the regular representation on $C^\infty(G)$ commutes with Δ, and therefore the eigenspaces of Δ (which are finite dimensional) are G-invariant. Hence, the eigenspace decomposition of D gives a decomposition of $L^2(G)$ into finite dimensional G-invariant subspaces. In other words, for compact Lie groups, the Peter-Weyl theorem (as we have stated it in 3.3.4) (and which, we recall, we deduced from the spectral theorem for compact operators), follows from the generalization of the spectral theorem for elliptic operators (6.5.1) to compact manifolds. Once again, if G is the circle, the G-invariant Laplace-Beltrami operator is $\Delta = d^2/d\theta^2$, and the eigenfunctions are $\{e^{in\theta} \mid n \in \mathbf{Z}\}$.

EXERCISES

6.1 Suppose $\Omega \subset \mathbf{R}^n$ is open and $\Omega = \bigcup_{i \in I} V_i$ is a union of open sets. If $T_1, T_2 \in \mathcal{D}'(\Omega)$ and $T_1 | V_i = T_2 | V_i$ for each i, show $T_1 = T_2$.

6.2 If $\Omega = \cup V_i$, $T \in \mathcal{D}'(\Omega)$, and for each i there is some $f_i \in L^1_{\mathrm{loc}}(V_i)$ such that $T | V_i = T_{f_i}$, show that $T = T_f$ where $f \in L^1_{\mathrm{loc}}(\Omega)$ and $f | V_i = f_i$ a.e.

6.3 Verify the assertion of Example 6.1.9(b).

6.4 If $f \in L^1(\mathbf{R}^n)$, show $f \in H^k(\mathbf{R}^n)$ for some $k \in \mathbf{Z}$.

6.5 If $U \subset V$, show that for any $k \in \mathbf{Z}$ there is a natural inclusion $H^k(U) \subset H^k(V)$.

6.6 If $\Omega = \cup V_i$ and $T_i \in \mathcal{D}'(V_i)$ such that for all i, j, $T_i | V_i \cap V_j = T_j | V_i \cap V_j$, show there is a unique $T \in \mathcal{D}'(\Omega)$ such that $T | V_i = T_i$.

6.7 Suppose $T \in \mathcal{D}'(\Omega)$ and $\partial_i T = 0$ for all i. If Ω is connected, show T is a constant function.

6.8 Suppose μ is a finite measure on $\mathbf{R}^n$. Show there is some $f \in L^1_{\mathrm{loc}}(\mathbf{R}^n)$ such that $\mu = Df$ (as distributions) where $D = \partial_1 \circ \cdots \circ \partial_n$. Hint: For $\varphi \in C^\infty_c(\mathbf{R}^n)$ write $\varphi(x) = \int_{I_x} D\varphi$, where $I_x = \prod_{i=1}^n (-\infty, x_i)$.

6.9 If $\Omega \subset \mathbf{R}^n$ is open and $\varphi : \Omega \to \Omega$ is a diffeomorphism, define $\varphi^* : \mathcal{D}'(\Omega) \to \mathcal{D}'(\Omega)$ to be the adjoint of the translation operator φ defined on $\mathcal{D}(\Omega)$. Find all $T \in \mathcal{D}'(\mathbf{R}^n)$ which are invariant under the action of $\mathbf{R}^n$ on itself by translation.

6.10 A distribution on $\mathbf{R}^n$ is called tempered if it is continuous with respect to the topology on $C^\infty_c(\mathbf{R}^n)$ as a subspace of the Schwartz space of rapidly decreasing functions (exercise 5.4). Show any $T \in \mathcal{D}'_c(\Omega)$ is tempered. Show any $f \in L^p(\mathbf{R}^n)$ $(1 \le p \le \infty)$ defines a tempered distribution. Give an example of a non-tempered distribution.

6.11 Let $T \in \mathcal{D}'(\Omega)$. Show there is a sequence $T_j \in \mathcal{D}'_c(\Omega)$ such that for every open $V \subset \Omega$ with $\overline{V} \subset \Omega$ and $\overline{V}$ compact, we have $T_j | V = T | V$ for all sufficiently large j.

6.12 Let D be an elliptic operator on Ω and $\lambda \in \mathbf{C}$. Show that $\{\varphi \in L^2(\Omega) \mid \varphi \in C^\infty(\Omega)$ and $D\varphi = \lambda\varphi\}$ is closed in $L^2(\Omega)$.

6.13 Suppose $\varphi \in C([a, b])$ and that φ is smooth on an open interval containing $[a, b]$. If $\varphi \in H^1((a, b))$, show $\varphi(a) = \varphi(b) = 0$.

6.14 If $\Omega \subset \mathbf{R}^n$ is a bounded open set and $u \in C^\infty(\Omega)$ is harmonic (i.e., $\Delta u = 0$), show $u \notin H^1(\Omega)$ unless u is identically 0.

6.15 a) If $\Omega \subset \mathbf{R}^n$ is bounded and $\lambda \in \mathbf{C}$, show there is some $f \in L^2(\Omega)$, $f \neq 0$, such that $\Delta f = \lambda f$.

b) Show $I - \Delta : H^2(\Omega) \to L^2(\Omega)$ has closed range, but that it is not surjective.

INDEX